T0292554

Industrial Espionage and Technical Surveillance Counter Measurers

Iosif Androulidakis
Fragkiskos – Emmanouil Kioupakis

Industrial Espionage and Technical Surveillance Counter Measurers

Iosif Androulidakis
Researcher
Pedini Ioannina, Greece

Fragkiskos – Emmanouil Kioupakis
Akmi Metropolitan College
Maroussi, Greece

ISBN 978-3-319-28665-5 ISBN 978-3-319-28666-2 (eBook)
DOI 10.1007/978-3-319-28666-2

Library of Congress Control Number: 2015960240

Springer Cham Heidelberg New York Dordrecht London

Printed on acid-free paper

Springer International Publishing AG Switzerland is part of Springer Science+Business Media (www.springer.com)

Preface

Industrial Espionage and Technical Surveillance Countermeasures

Undoubtedly, the protection of intellectual property and confidential data in the information era is vital, not only for the evolution of a company but also for just keeping it "alive" in the modern, extremely competitive market. Unlike the "traditional" espionage that is conducted for national security purposes, the practice of collecting confidential information without authorization from its owner for commercial or financial purposes is called industrial, corporate, commercial, or economic espionage. Despite the fact that the techniques that information "spies" use and the industrial espionage incidents themselves look like they have stemmed out of a movie, the truth is that it is a real, existing problem, more severe than ever. This book tries to shed more light on this subject and inform the readers about the risks involved as well as about the ways they can protect themselves and their companies.

In the first chapter, we analyze the phenomenon, giving statistics and describing real incidents in the past recent years. The major contribution of this book is the extensive description of the equipment and techniques used by potential eavesdroppers in order for the right precautionary measures to be taken. In that manner, the second chapter is dedicated to the interception of ambient conversations, the third one to data interceptions, the fourth one to intercepting fixed line communications, and the fifth is dedicated to intercepting mobile communications. These chapters are accompanied with a rich collection of photos while also including practical advice for reducing the associated risks. The sixth chapter presents an example of a malware that can be used for industrial espionage or as a personal spyware and a way to protect from it. The book concludes with the seventh chapter which includes an equally extensive description of the countermeasures that can be implemented for protecting intellectual property both on a legal and on a technical level. It also deals with detecting interceptions using specialized equipment. The same chapter further presents guidelines and methodologies for audit controls and also for selecting the appropriate contractor for a successful countersurveillance project.

Athens
November, 2015

Dr. Iosif Androulidakis
Fragkiskos – Emmanouil Kioupakis

Acknowledgements

At this point, I would like to thank my family for their love and moral support as well as my professors for their guidance during my studies in Greece and abroad. Authoring a book is a very demanding process. Thanks to the excellent collaboration with Fragkiskos, everything went smoothly and flawless. I would like to thank him very much and also thank the countless information and telecommunication security researchers and professionals that I have met so far. Having completed dozens of projects and given more than 120 relevant speeches in conferences and events in 20 countries, the experience gained from cooperating and discussing with them was valuable and crucial for authoring the book that you now hold in your hands. I hope you enjoy reading and also hope that with the information in the following pages you will be able to protect yourselves from the interception of sensitive information, commercial or not.

–Iosif Androulidakis

In my turn I would first like to thank Iosif very much for giving me the opportunity to work with him for authoring this book. As he is a well-esteemed researcher, the collaboration with him offered me valuable knowledge since he was more of a mentor to me rather than simply a colleague. This opportunity also enabled me to accomplish objectives that otherwise I couldn't have accomplished so successfully and early in my research career. I would also like to thank my professor who also was my supervisor on my Master's dissertation as he was the one who introduced me to academic research, offering his help and support. This introduction to academic research was also what led to meeting Iosif, thus being a turning point for me, especially in regard to this book being coauthored by me. Last but not least, I would like to thank my family for all the necessary support that I was given throughout my studies.

–Fragkiskos – Emmanouil Kioupakis

Contents

About the Authors

Iosif Androulidakis has an active presence in the ICT security and countersurveillance field. He has authored more than 90 publications (including six books) and has presented more than 120 talks and lectures in international conferences and seminars in 20 countries, and he has participated in many security projects.

Holding two Ph.D.s his research interests focus on security issues in telecommunication systems where he has more than 20 years of experience. Part of his research has led to the granting of five patents. During his career, he collaborated with telecom operators, national police cybercrime departments in many countries, the European Police Academy (CEPOL), the European Public Law Center, the Southeastern Europe Telecommunications and Informatics Research Institute, universities and research centers, vocational training institutes, and the media and private security consulting firms.

Dr. Androulidakis has also acted as a reviewer in an extended array of scientific conferences and journals, as a programme committee member in 26 conferences, and as a chairman in 8 conference sessions. Finally, he is a certified ISO9001 (Quality Management System) and ISO27001 (Information Security Management System) auditor and consultant.

Fragkiskos – Emmanouil Kioupakis obtained his B.Sc. in computing in 2013 and his M.Sc. in information security and computer forensics in 2014. He has published a number of research papers on the subjects of Android malware and information security rating methodologies and been a speaker in conferences regarding the same subjects. He has been a network engineer working at various ISPs in Greece for several years. His interests include the subjects of information security, information security management and rating methodologies, optical network technologies, and next-generation access networks.

Chapter 1
Industrial Espionage

1.1 Introduction

In the introductory chapter of the book, we will be analyzing the subject of industrial espionage, providing introductory elements, statistics relevant to the problem the entities involved, as well as real cases of industrial espionage. Despite the fact that the cases we will be examining seem to be coming out of James Bond movies, industrial espionage today is more widespread than ever, threatening the majority of companies and organizations.

1.2 Espionage in General

Espionage is the practice of collecting confidential information (mostly of governmental nature) without the authorization of its owner. The USA defines espionage toward itself as "The act of obtaining, delivering, transmitting, communicating, or receiving information about the national defense with an intent, or reason to believe, that the information may be used to the injury of the United States or to the advantage of any foreign nation." Unlike other forms of information gathering that takes place remotely (e.g., penetrating computer networks), traditional espionage mostly relates to accessing the physical site location where the desired information is kept, as well as the people that handle that specific information. In the second case, spies will try to extract information from the individuals that are authorized to know it, using either tricks or equipment such as interception setups and devices.

I. Androulidakis, F.–E. Kioupakis, *Industrial Espionage and Technical Surveillance Counter Measurers*, DOI 10.1007/978-3-319-28666-2_1

1.3 Industrial, Corporate, Commercial, or Economic Espionage

Leaving aside the governmental intelligence, espionage that is committed for commercial and economic purposes is called industrial, corporate, commercial, or economic espionage. Among other things, it includes stealing trade secrets, bribing, blackmailing, as well as using advanced electronic means of surveillance and interception that will be further analyzed in the subsequent chapters. At this point it is worth mentioning that although commercial in form, even a nation can become a victim of this type of espionage (e.g., during a public tender competition for the purchase of high-value equipment/services).

The importance of industrial espionage is so great that Pierre Marion, the former Director of the French General Directorate of External Security (DGSE-Direction générale de la sécurité extérieure), publicly stated that: "This espionage activity is an essential way for France to keep abreast of international commerce and technology. Of course, it was directed against the United States as well as others. You must remember that while we are allies in defense matters, we are also economic competitors in the world." On the other side of the Atlantic, the former Director of FBI, Louis Freeh, stated that: "Economic espionage is the greatest threat to our national security since the Cold War."

Especially regarding the relationship with China, the United States-China Economic and Security Review Commission, in late 2007, stated that: "Chinese espionage in the United States, which now comprises the single greatest threat to U.S. technology, is straining the U.S. counterintelligence establishment. This illicit activity significantly contributes to China's military modernization and acquisition of new capabilities" and also that "In order to slow or stop the outflow of protected U.S. technologies and manufacturing expertise to China, the Commission recommends that Congress assess the adequacy of and, if needed, provide additional funding for U.S. export control enforcement and counterintelligence efforts, specifically those tasked with detecting and preventing illicit technology transfers to China and Chinese state-sponsored industrial espionage operations."

1.4 Business: Competitive Intelligence

Besides the "illegal" forms of espionage, there also exist a number of legal "espionage" techniques and methodologies (with varying degrees of "ethics") that as a whole can be referred to as business or competitive intelligence. Indeed, the interested individuals can gather information from a wide variety of means in a totally legal manner. Examples include newspapers, balance sheets, submission of patents, publications in scientific journals, information from court trials, various advertising data, websites, etc.

Moreover, it has become routinely act among competitors to try and “mine” information during exhibitions and conferences. Researchers and sales staff attend this kind of exhibitions to stay up to date with technology and evolution in their market but also to promote their company’s products. Often though, during promotion, they reveal more information than they should be revealing in the first place. The competitors use professionals and experts at mining information, to extract as more as possible important data that the employees should not have shared. These experts often present themselves as potential clients or scientists. With advanced techniques and proper training, they extract information from willing and enthusiastic employees who don’t suspect the “trap.” This way, it is not immediately apparent that the company is actually revealing all its secrets by itself allowing its competitors to reach to valuable conclusions regarding their status and their products. Furthermore, the discovery of such a data leakage can take too long, rendering it late to mitigate the consequences of the information leak.

Governments typically show particular interest for their own citizens that have worked in high-profile jobs in companies and organizations of other countries. It is only logical that an individual will show more loyalty to his/her own country rather than to a foreign company. Therefore, if asked to reveal information from his previous employee, for the benefit of national interest, it is very possible that he/she will accept to do so. On an even more advanced level of such cases, a government itself may intercede to aid a citizen, in any possible means, to be hired in a company or organization of a foreign country in order to use that individual later as a spy.

On the market domain, high-level executives moving between companies practically equals to the transfer of knowledge between competitors. Even involuntary, employees use expertise gained from their previous employers in their everyday work. They also, inevitably, bring with them part of their clientele.

Joint ventures are also an important factor of leakage of secrets since in order to expand the level of technology and the production of new products, it is necessary to reveal information about the state of the art currently being used. In the same sense, a country may ask for “reciprocal” benefits for allowing a foreign company to operate in its territory. These “reciprocal” benefits may include staff training as well as technology and information transfer. Finally let’s don’t forget the most effective practice: the buyout of a company from another one, which automatically results in transfer of technology. Of course, this is the most expensive way to get to the “secrets,” so practically there is plenty of room for classical corporate espionage operations.

1.5 The Problem

Industrial espionage is more common in high-tech industries like electronics, automotive, pharmaceutical, chemistry, biology, aerospace, and energy. In these highly competitive industries, once data leakage occurs, there is very little that can be done to limit the impact, while losses can be overwhelming. The problem is intensified due to many factors that we will briefly represent in Table 1.1.

Table 1.1 Factors that aggravate the problem of industrial espionage

There are huge amounts of money at stake, while the "paychecks" of the individuals involved can be equally big. In times of financial crisis that, among other things, mergers and buyouts are taking place, the value of information is becoming even more important
Employees (and especially administrators) of Information Technology departments have access to systems and servers that host sensitive information, source code, credit card details, blueprints, etc.
Even worse, in many cases, the employees aren't informed about the criticality of the information that they handle, resulting in inadequate protection of it
There are dozens of ways that data leakage can occur, including floppy disks, USB sticks, CDs/DVDs, smartphones, and MP3/4 players but also by using network services (e-mail, FTP, VPN access, etc.) In the well-known case of WikiLeaks, soldier Bradley Manning very characteristically revealed that he has been copying classified data on rewritable CDs, freely passing through security checks by stating that they contained music by Lady Gaga!
The lack of security certainly renders the situation worse, but on the other hand, technical means of security may be overused. That may create a false sense of security that in turn can lead to slackness and over-relaxation
At the same time, restrictions due to security policies hamper the daily routine of employees, and it is for this reason that it is a common practice for those restrictions to be disregarded or overridden
As previously mentioned, there are also cases of sensitive information being revealed inadvertently. Scientists during speeches often reveal sensitive information due to their enthusiasm with no intention to harm the company. Even the company itself may make similar mistakes by trying to advertise a product in the best possible way
The modern business environment dictates limiting costs on all levels. Information needs to be transmitted quickly between companies themselves but also among their partners, aligning with the fast rates that markets operate. The problem expands outside the strict company limits and even outside the country in the case of multinational companies. The secure exchange of information between the company, the partners, and the suppliers has become a very challenging process
In the same manner, the assignment of contracts to other companies (outsourcing) that are usually based in countries outside Europe and USA can be a big "headache"
Let's not forget that new products and new technologies also create new threats. Small research and development companies with high degree of innovation often focus primarily on research, ignoring (or not being able to afford) or protecting their intellectual property

1.6 The Human Factor

There are many "players" involved in the "game" of corporate espionage. Foreign governments, competitors, and small companies that do not have the necessary resources to develop new technologies themselves but also individuals with revenge or profit motives are the most known ones.

To accomplish their plans, they use professional information brokers, dissatisfied employees, employees that are hired with the only intention to conduct espionage, black hat hackers, and spies. The latter category may include a wide variety of individuals such as university students, immigrants, translators, journalists, tour guides, etc. With the role and the task that they are assigned to complete, these individuals can either be characterized as "heroes" or "traitors," depending on the point of view!

Within companies, the human factor remains to be one of the most basic ones concerning industrial espionage. Employees (and especially administrators) of Information Technology departments have access to systems and servers that host sensitive information, source code, credit card details, blueprints, and many other company secrets.

An espionage option with a longer time frame horizon of action is placing an employee spy at the competitive company. He/she maybe hired in any operational level of the company, providing valuable information from any possible position. Even at the lowest possible position such as a janitor or as a clerk for outside jobs, they can prove to be more valuable at collecting sensitive information than even a high-level executive. In some cases, this "internal" point of contact may be found at the face of a dissatisfied employee that desires to get revenge against the company he/she works in (and mostly against his/hers supervisor). Such an employee would be more than happy to provide information to the rival company in order to damage the company that "betrayed" him/her.

As already mentioned before, "spies" will try to extract information from specific personnel that has access to that information, by using some kind of trick. However, this may also happen in a more direct way. The simplest method is trying to bribe the authorized individual, especially in case that individual is known to be facing financial problems. This highlights the importance of background checks that spies perform in order to pick their target.

A more aggressive way is blackmailing which may also be very effective since employees are only humans after all. The "victim" may also be framed by the perpetrators, and the evidence may be a result of actions that are organized beforehand.

At the same time, experts at mining information (using legitimate, semi-legitimate, or completely illegitimate means) exploit every public source of information as well as employees that are more "talkative" than they should be. These experts use social engineering tactics, pretending to be someone else in order to extract information and to convince others to do something that they would not do under normal circumstances. According to Wikipedia, social engineering in the context of information security refers to psychological manipulation of people into performing actions or divulging confidential information. All social engineering techniques are based on specific attributes of human decision-making known as cognitive biases. These biases are exploited in various combinations to create attack techniques.

1.7 Statistics

Many studies have been conducted to show the extent of the problem. The Information Security Breaches Survey 2014 report from PricewaterhouseCoopers mentions that 55 % of large organizations and 14 % of small businesses in the UK encountered loss or leakage of confidential information from their employees in 2014 (Fig. 1.1).

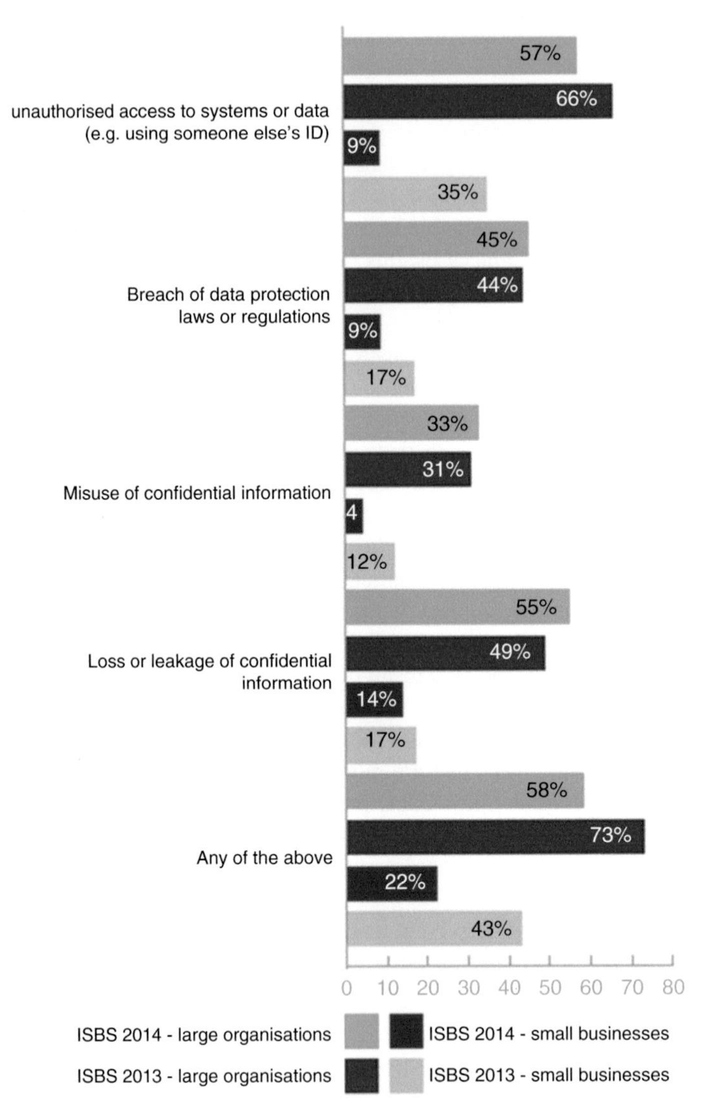

Fig. 1.1 Staff-related information security issues, Information Security Breaches Survey 2014, PricewaterhouseCoopers

The Global Information Security Survey study from Ernst and Young for 2014 indicates that 44 % of the companies worldwide stated that cyberattacks to steal intellectual property or data are the first or second threat to increase their risk

Which threats and vulnerabilities have most increased your risk exposure over the last 12 months?

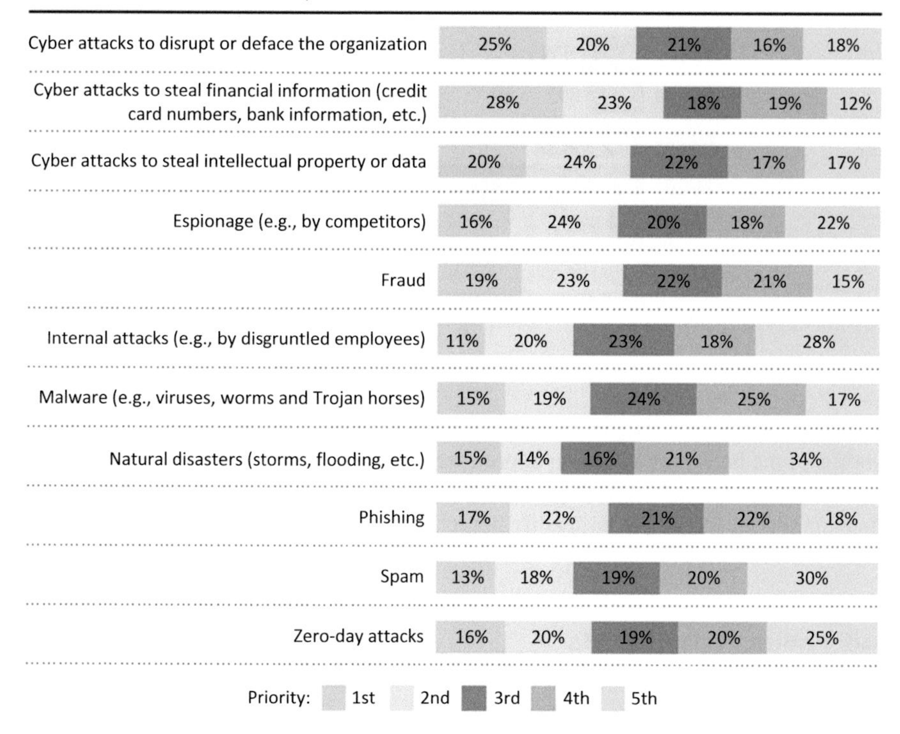

Fig. 1.2 Threats and vulnerabilities that increase risk exposure, Ernst & Young, Global Information Security Survey, 2014

exposure. According to the same study, 40 % of the companies worldwide classify industrial espionage as the first or second threat for risk exposure, respectively, while internal attacks from employees rank high in the study with 31 % (Fig. 1.2).

Important data revealed in the Data Breach Investigations Report of Verizon for 2014 shows that 62 % of companies worldwide are victims of industrial espionage and needed months to discover the incident. A limited, but worth mentioning, percentage of 5 % of the victims of industrial espionage needed years to discover the incident (Fig. 1.3). These facts make the overall impact of industrial espionage considerably bigger.

It is quite difficult to estimate the financial losses for companies resulting from stealing confidential information, since these estimations are based on complicated and numerous factors. However, McAfee estimated in the Economic Impact of Cybercrime and Cyber Espionage study in 2013 that financial losses range from 120 to 280 billion dollars a year just in the USA, while 75 % of those losses are due to loss of intellectual property material.

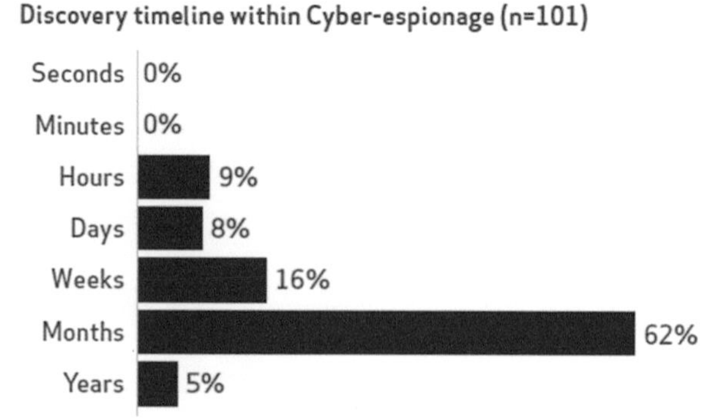

Fig. 1.3 Discovery timeline for cases of industrial cyber espionage, Data Breach Investigations Report, Verizon, 2014

It is also important to note that studies of this nature are usually based on extrapolation of the known cases. Many more cases never go public in order not to damage the company's reputation and the investors' and public's trust. Therefore, the actual financial losses may be significantly bigger.

1.8 Some Actual Facts

In the next two paragraphs, we cite some of the most interesting incidents and cases of the recent years. In fact we can make two simple assumptions: first, the cases that go public are only the tip of the iceberg, and second, many more spies manage to evade than those that get caught and end up in courthouses.

The subject of espionage (industrial and nonindustrial) has recently became popular to the general public from incidents regarding WikiLeaks and the global surveillance disclosures that kept the media busy for a long time. Besides that, the list of industrial espionage incidents is a long one, and among the victims are giants like Intel, Unilever, Cisco, Bristol-Myers Squibb, Gillette, Kodak, and others. In one of the most costly cases, Lockheed Martin suffered losses of one billion dollars due to intellectual property theft.

In 2013 AMD filed a lawsuit against a number of former employees accusing them for downloading over 100,000 sensitive documents onto external drives before joining Nvidia. AMD said it forensically analyzed the employees' computers and found evidence of intentional access to AMD's secure computers. In the lawsuit AMD accused them for misappropriation of trade secrets, violation of unfair competition laws, computer fraud, breach of employee's duty of loyalty, breach of contract, and conspiracy for the 6-month-long theft of confidential documents before accepting a position at Nvidia.

In 2013, the networks of two banks and three media companies of South Korea were paralyzed during a well-organized cyberattack. Despite the fact that potential sources of that attack were not named, there were very specific suspicions. Furthermore, after the attack, government sources of South Korea expressed their concern regarding future simultaneous attacks against infrastructures of energy, transport, telecommunications, as well as military and government infrastructures.

In 2011, the information security giant RSA was attacked with the usage of an advanced persistent threat (APT), with the purpose of intercepting encryption keys for RSA's product SecurID that RSA supplies to US government organizations, intelligence agencies, defense contractors, and Fortune 100 companies to use for accessing encrypted systems.

Sony's PlayStation Network (PSN) was also attacked in 2011 by, among others, DDoS attacks, resulting in the leakage of data regarding 100,000,000 users of PSN. Those data included users' names and accounts, postal addresses, and credit card data.

The banking group of Citi, discovered an attack by chance, during a routine check in 2011 that resulted in data theft regarding 200,000 Citi's clients. Citi suffered losses of two million dollars, being forced to reissue a big number of credit cards due to that attack.

Pirelli was accused in 2010 for telephone interceptions conducted by Telecom Italia, which is one of the subsidiaries of Pirelli. Even though those interceptions were targeted against politicians, it is believed that same practices were used to conduct industrial espionage against some of Pirelli's competitors such as Michelin.

Nortel, a big telecommunications equipment manufacturer, was bankrupt in 2009. Industrial espionage against Nortel including the theft of important confidential data took place for 10 whole years without being discovered. This can very well have played an important role.

GhostNet was an extended surveillance system that was discovered by Canadian researchers at the University of Toronto. With targeted cyberattacks, the system had access to thousands of computers belonging to government organizations and agencies, allowing the attackers to copy information and transfer it to servers in China.

The Stuxnet worm is an example of advanced software that managed to delay the development of Iran's nuclear program, exploiting vulnerabilities in specific systems of industrial control (PLCs). The sophistication of Stuxnet and its' targeted operation resembles "governmental" involvement.

"Operation Aurora," as it became known to public, involved theft of intellectual property from Google China and access to e-mail accounts belonging to activists. The investigations that followed related the attacks with two Chinese universities, one of which was closely related to the army. This incident resulted in the termination of Google operations in China, on March of 2010.

In 2007 Ferrari Formula One team accused a senior McLaren Formula One engineer and former Ferrari team employee for passing confidential information to McLaren. At the same time there were allegations that Renault Formula One team was also being passed confidential information from McLaren. That confidential information allegedly included technical specifications for Formula One cars and other technical design information. Following the investigation from International Automobile Federation (FIA), the McLaren Formula One team was excluded from the 2007 Constructor's Championship and also resulted in record-breaking fine of 100 million dollars.

During a rare case of cooperation between competitors, a Coca-Cola employee and his two accomplices were arrested in Atlanta for stealing intellectual property (recipe of a new product) and trying to sell that information to Coca-Cola's competitor PepsiCo. PepsiCo informed Coca-Cola and the FBI and the cooperation led to the arrest of the perpetrators in 2006.

In 2005 a Netgear employee, after being offered a position at Broadcom and before leaving Netgear, downloaded dozens of trade secret files from Marvell Semiconductor. Marvell is a business partner of Netgear, providing chips for Netgear's products and a direct competitor to Broadcom. Due to his position at Netgear, the employee had access to Marvell's secure database information, and before leaving Netgear, he was downloading files from the database to a Broadcom issued laptop for 9 days.

In 2004, there were rumors that conversations of Marks and Spencer's CEO were leaked. When Marks and Spencer contacted two top risk management and consulting companies, both refused to handle the case due to conflict of interest. Obviously others have already assigned to those two companies to examine the case and to potentially reveal who had the conversations at his disposal!

1.9 Summary

Apart from the cases mentioned above and many other cases that went public, it is 100 % certain that every day, dozens of industrial espionage incidents take place that will probably never be revealed. This book aims to inform readers and help them protect themselves so they will not be listed in the long list of victims of espionage. In the next five chapters of this book, we will present the techniques and methods used by the entities involved in espionage for intercepting conversations, data, and telecommunications. This book will be completed with the seventh chapter that analyzes the countermeasures for industrial espionage. We will conclude the introduction with a known joke: "An employee shows up in a bar frustrated asking the other customers: I lost some intellectual property here last night. Anybody remember what the hell I was talking about?"

Chapter 2
Intercepting Ambient Conversations

2.1 Introduction

Continuing from the previous introductory chapter regarding industrial espionage, in this chapter we will focus our attention on the technical means, methods, and equipment that "spies" use for intercepting ambient conversations from the room/ place that the victim is located. In the next chapters we will continue the analysis with the interception of data and the interception of telecommunications.

2.2 Interception Devices

As we have previously mentioned, information isn't stored only on computers, but it is also transmitted from individuals via conversations and phone calls. Thus, there is a series of different interception devices that aim to record audio and/or video and possibly transmit this information to a different location (these devices are collectively known as "bugs"). The rapid development of technology gives access to a plethora of electronic devices and parts used for interceptions, available in sizes small enough to make their visual trace almost impossible for the non-trained eye. Figure 2.1 depicts one of these devices.

Such devices, thanks to their small size, can be hidden in every possible location one can imagine. They can also be built-in or embedded in other objects or even "worn" on someone that was able to penetrate the area where the conversation to be intercepted takes place. Figure 2.2 shows a man's belt that is equipped with a built-in microcamera. The very small hole that barely can be seen is the opening for the microcamera's lens. There are also various different forms of cameras and microphones that are housed in sunglasses, ties, coat buttons, etc.

I. Androulidakis, F.–E. Kioupakis, *Industrial Espionage and Technical Surveillance Counter Measurers*, DOI 10.1007/978-3-319-28666-2_2

Fig. 2.1 Wireless setup for intercepting room conversations

Fig. 2.2 Belt with built-in microcamera

"Bugs" can be used to monitor telephones, conference rooms, and every possible location company's executives may be having a discussion. Such kinds of devices have been found even hidden in airplane seats of business class and also in fine dining restaurants where executives were discussing important deals during business lunches.

A fairly common practice also involves sending someone a gift. Like a modern "Trojan Horse," the gift will be "bugged" with an interception device. There are dozens of such products as shown in Fig. 2.3 (desk clock, wall clock, calculator, etc.). A major limitation of "bugs" is the energy aspect. Battery-powered ones have certainly a limited operational time amount. Imagine, however, the case where the "Trojan Horse" gift itself is a mains-operated electronic device (e.g., a nice table lamp): the power source is now ensured, allowing the interception device to operate over a long period of time! In addition to that, the mains or telephony plug

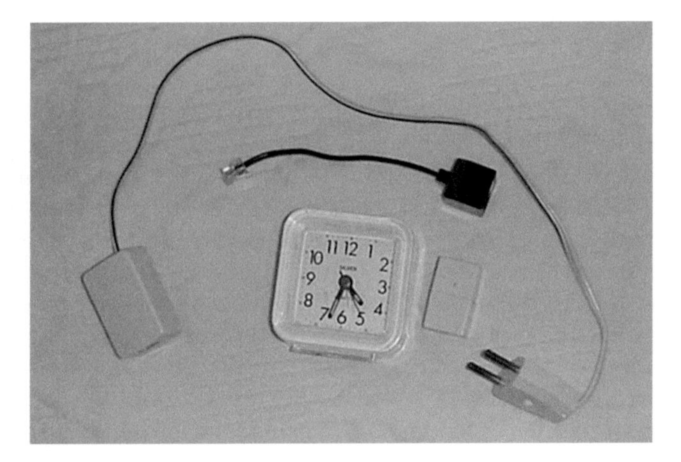

Fig. 2.3 Interception setups housed in various objects

Fig. 2.4 Digital voice micro-recorder

itself allows enough space to accommodate such a "guest" while at same time providing plenty of power.

Regarding their operation, these setups may either transmit or store the intercepted data. A simple digital voice recorder (or specialized devices as the ones shown in Fig. 2.4) can record dozens of hours of conversations. Furthermore, it can even be activated upon voice detection in order to serve the spy for many days before eventually running out of battery power. Of course, it is imperative to ensure the removal of the device from the room it is located in order to retrieve the data recorded. Thus, usage of a digital voice recorder requires a second or even more "visits" to the room to be "bugged," and this can prove to be difficult. Transmitting devices that will be analyzed shortly below overcome this limitation.

It is also possible to intercept wireless transmissions (from Wi-Fi networks but also from cellphones and cordless phones). We will refer to both Wi-Fi and cellphone and cordless phone interceptions in the following chapters.

2.3 Transmitting Setups

Video and audio recorders are "passive" in nature, in the sense that they do not transmit any information. Therefore, it can be more difficult to be located. Transmitting devices, however, have an important advantage. The intercepted information is immediately available and does not require any second visit to the area in order to retrieve the bugging device. This is especially important in cases where rapid decisions must be made and the information must be available in real time. On the other hand, given their limited size and the energy efficiency required, their range is limited. This is why there often exists a repeater (or a monitoring station) in an office close by to the intercepted one or in a car parked outside the area that the interception is conducted. The repeater, hosted in a place where more room available receives the initial transmission, amplifies it and retransmits it in a great distance, so that the spy can get access. Indeed, the original minitransmitter might be hidden in the desk of the CEO, while a bigger size, mains-operated repeater can be placed in an adjacent office (even in another floor).

A very effective way to extend the transmission range is by using a modified cellphone that the spy has carefully hidden in the room. In this case, the cellphone automatically and silently answers a predetermined phone call, and it transmits the conversations to the other end. Therefore, there is no need to use a repeater anymore, since the intercepted material can be received even from another country, thousands of miles away. This practice has the disadvantage of the relatively bigger size of the cellphone, but even so, the cellphone may be deceitfully hidden inside a larger "innocent" device. Of course, there already exist specific devices available in the market to be exclusively used for such cases with a significantly smaller footprint. Figure 2.5 depicts such a device, equipped with a SIM card. These devices are essentially a micro-cellphone (without a screen and keyboard) with a powerful microphone. They require a plain mobile telephony SIM card, and they can be called as a normal telephone. As already mentioned, when called, they silently take

Fig. 2.5 Interception device transmitting via the cellular network

the call and switch the microphone on, so that the spy can listen to the ambient sounds. The major disadvantage of such an implementation is the very limited operating time (a few hours). Should, however, an external energy source is provided, then they can operate for much more time.

As already mentioned, transmitting devices can be detected since they are "active" in nature. Their transmission (and therefore their location) can be detected by using specialized equipment (such as spectrum analyzers or field meters) as we will be discussing in the respective chapter. This is why a more elaborate (and expensive) alternative exists. This device records conversations similarly to a digital voice recorder but does not transmit them immediately. Instead, after collecting enough data, it converts them to a digital signal, then compresses it, and transmits it, possibly encrypted. This way, the signal transmission is not constant (as is the case with normal transmitters), making the detection incredibly difficult (it is required to keep using a detector in the room for many hours). An equally difficult to detect device uses a spread spectrum transmitter that uses the direct-sequence spread spectrum (DSSS) or the frequency hopping spread spectrum (FHSS) technology. In both cases it is very hard to detect such transmissions. In the first case, the difficulty derives from the fact that the transmission is low power but in wide frequency spectrum so it is not easily distinguished from the electromagnetic noise of the environment, whereas in the second case, the transmission rapidly switches between different frequencies.

2.4 Microphones

Microphones are the core elements of every ambient audio-intercepting device. Apart from their use in devices, they can also be used standalone. Installing a high-sensitivity microphone (Fig. 2.6) and wiring it to an adjoining office is a simple solution. Of course, the wiring part is the most revealing one and can be spotted. The usage of a conductive pen (or conductive paint) is impressive.

Fig. 2.6 High-sensitivity microphone

Fig. 2.7 Wall microphone

Fig. 2.8 Parabolic microphone

This paint (usually based on silver) is commonly used to repair broken car rear window heating elements. The trail of the pen creates a conductive film as in an electrical conductor of practically zero thickness. It therefore allows connecting a microphone, across the length of a wall. Repainting the wall to cover the trail of the pen, one ends up with a completely stealth interconnection. Interception from an adjoining room sharing a common wall is equally effective using a stethoscope or a simple water glass. For professionals, there are in fact advanced electronic "stethoscopes" (wall microphones) to be used for such purposes (Fig. 2.7).

Another category of microphones is directional microphones, usually parabolic ones (Fig. 2.8). These microphones can pick up sounds from a great distance. However, when operated in high-level noise environments such as cities, their effectiveness is limited. On the contrary, they are quite popular among nature lovers who are able to hear various birds and animals from a distance in quite places such as forests.

Finally, sound can propagate in a natural way to nearby areas, without the use of equipment or microphones, through a hole on the wall, air-conditioning ducts, chimneys, heating pipes, etc. It is indeed remarkable how we focus our attention on high-tech electronics, while construction flaws such as thin walls or lack of audio insulation are more than enough for a "leakage."

2.5 Exotic Tools

More advanced "tools" include a modified incandescent light bulb whose lighting intensity is modulated by the sounds of the environment, producing an unnoticeable to the human eye flickering. With the appropriate equipment, these minor changes in the light intensity are converted back to audio. A similar real-world analogy is that of the disco lights that light according to the music. In that case, the effect of course is visible to the naked eye. Needless to mention that there must be a line of sight connection between the target and the spy.

Another line of sight technique can be described from the espionage case in the US Ambassador's office in Moscow, in 1945. Back then, the Soviets gave the US Ambassador a carved wood plaque of the Great Seal of the United States as a gift and a "gesture of friendship." However, this gift was nothing more but a covert listening device. The most impressive aspect of the case is the bug's design. Designed by Leon Theremin, the device was passive and only activated when it was radiated by a signal at a precise frequency (it is therefore considered as a predecessor to RFID). That could also be done from a distance, making it perfect for that use.

The device was equipped with a membrane in a cavity that resonated at the frequency of the beamed signal. When that signal was transmitted, the device would be modulating the sound it received through the wood panel in front of it and would transmit it at a different and higher frequency than the beam used to activate it. The bug's transmitting signal could then picked up and demodulated to reveal the context of conversations taking place in the Ambassador's residential study.

Even though it had an antenna to transmit the intercepted conversations, its design made it very difficult to be discovered. Being completely passive, it had to power source, and it was only transmitting signals when it was resonated by the beaming signal at 330 MHz as it is said.

It took quite a while till the device was discovered by accident in 1952. A British radio operator picked up American conversations on an open radio channel, obviously at the same time as the Soviets were beaming signal to activate the device.

Another technique, usually portrayed in movies, is to intercept sound from the subtle vibrations on window glasses. This movement, measured in microns of a meter, changes the reflection angle of an invisible infrared laser beam whose beam can in turn be demodulated into sound. Again, in this case, every noise, such as vehicles passing by, dramatically decreases the system's efficiency.

Since we mentioned vehicles, even a car can possibly be "bugged," both for intercepting sound as well as tracking its' position with a GPS-based device. The list of "exotic" devices also includes devices used to intercept fax messages.

It is also possible to conduct a passive interception without any kind of intervention. In (forgotten by now) cathode-ray tube (CRT) monitors it was possible to remotely reconstruct the signal due to electromagnetic transmissions (Van Eck-TEMPEST effect). Based on the same principle, signal interception from wires is possible, through induction. In Fig. 2.9 a special inductive coil is shown which intercepts telephone signals without breaking the circuit, just by picking up the electromagnetic fluctuations. The basic principle here is that according to

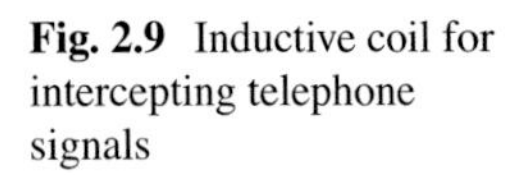

Fig. 2.9 Inductive coil for intercepting telephone signals

Maxwell's equations, a varying electric field produces a magnetic field and vice versa. Physical audio in electronic devices is transformed to a varying electrical current that produces a magnetic field. So, the induction coil picks the magnetic field, transforms it back to current, and drives an amplifier to allowing overhearing the initial audio.

2.6 Systems for Enforcing Secrecy of Communications

Having described various interception and bugging devices, we will describe some devices that aim to "neutralize" their presence. A conference room or an office can be equipped with the systems described in the following paragraphs to avoid the interception of confidential information during meetings.

A cellphone jammer (Fig. 2.10) covers all cellular bands to protect against hidden cellphones that act as interception devices. This tool is simply a transmitter that transmits along the whole frequency band mobile phones operate in. By masking the legitimate mobile telephony signals, this device renders mobile phone service inoperable in the nearby area. Depending on the power of the device, the jamming area can be extended to miles around it. It must be noted that in most places the use of cellphone jammers is illegal, because apart from the fact that interfere with the provider's network, they can also prohibit an emergency call from being placed.

Wireless camera and Wi-Fi network jammers (most wireless cameras operate on the same 2.4 GHz band as Wi-Fi networks do). Employing the same technique as the cellphone jammer, they operate in the frequency band where most wireless cameras operate, effectively blocking their transmission.

Fig. 2.10 Cellphone jammer

Analog and digital recorder jammers involve a device that prevents recording of conversations in a specific room, neutralizing the microphones of recorders present there. There are two basic categories of such devices. The simpler form looks like the one shown in Fig. 2.11 and creates white noise (random frequency and amplitude sound similar to the noise that an analog TV makes when is not tuned on a channel). This noise gets superimposed to the normal conversation. The combined sound recorded by the intercepting setup is very difficult to understand. Apparently, this sound is also audible from the participants, thus making it a quite annoying solution. More advanced devices like the one shown in Fig. 2.12 produce inaudible frequencies in the ultrasound band that have the same result. Although not necessary per se, in order to be unnoticeable, they can be housed inside an "innocent" device such as a clock.

Excellent audio interception security is provided using a special intercom system. The simplest form of consists of a helmet with a microphone and headphones that is connected to a same one used by the other participant. Voice is amplified from the microphone and transmitted to the headphones of the other participant,

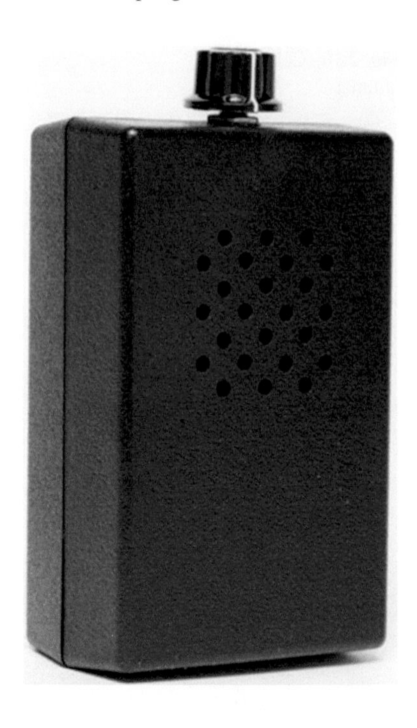

Fig. 2.11 White noise jammer

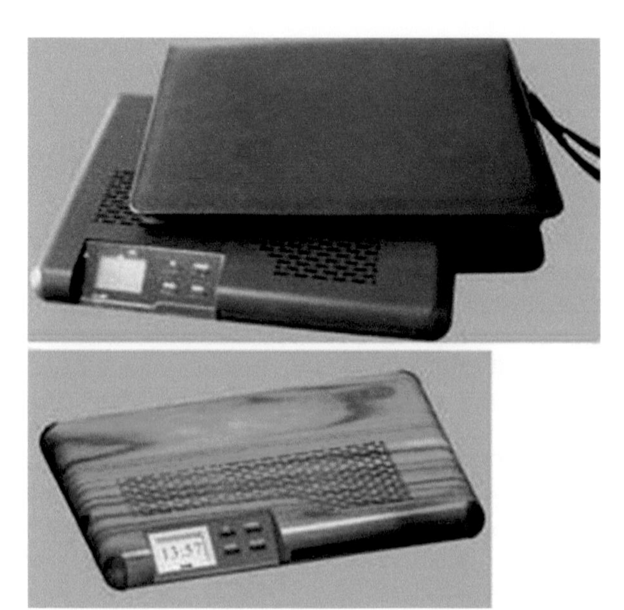

Fig. 2.12 Ultrasound jammers

while the helmet itself does not allow sound to be audible outside of it. Needless to say that it is not very practical and not many executives would like wearing it.

Having already secured that there are no interception devices present in the room, then telephone communications need to be protected as well. This can be made possible by using crypto devices (both for cellphones and landlines) that, as the name suggests, encrypt the telephone calls.

2.7 Summary

It is obvious that the devices portrayed in spy movies that were merely beyond imagination a few decades ago (James Bond, Mission Impossible, etc.) have become a reality. The most worrying fact is that nowadays their cost is very low (even 20€ are enough for a fully functional device), and they are relatively easy to find in the market. Detection on the other hand requires much more expensive equipment and expertise as we will discuss in the corresponding chapter, especially in the presence of professional and advanced interception devices. Continuing on to the next chapter, we will examine interception of computer data.

Chapter 3
Interception of Computer Data

3.1 Introduction

The "source" of data in the modern era is undisputedly computers. Whether standalone or networked, computers store and process information that can always be useful in cases of industrial espionage. The full description of the techniques and threats that may compromise the confidentiality of computer data, as well as the applicable countermeasures, is so extensive that cannot be included in a few pages of a single chapter, since it is the subject of the information security science. Moreover, the primary purpose of this book is to fill the gap in the available literature regarding industrial espionage and protection against telecommunication interceptions. In any case, for reasons of completeness, we will examine to a reasonable extent the software and hardware that is used for intercepting computer data.

3.2 Hardware for Intercepting Computer Data

Starting from the least obvious ones, from the hardware perspective we meet a very sophisticated circuit. It connects in-line to the computer's keyboard (in fact forming a cable extension) and records in its internal memory every keystroke of the user. It is called a "hardware key logger" to differentiate from similar purpose software. Such a device is shown in Fig. 3.1 and can be connected between the computer and the keyboard in seconds. There is a version for older PS/2 keyboards and a version for USB ones. After a few hours or days, the device is removed as quickly and as easily as it was installed and gets connected to the "spy's" computer. Using a predefined password, the contents of the keystrokes that were logged are displayed in the relevant application. Needless to say that all passwords that they were typed in, all documents, spreadsheets, and e-mails that were composed are now in the hands of the criminal. Even the spelling errors are displayed, since the "backspace"

I. Androulidakis, F.–E. Kioupakis, *Industrial Espionage and Technical Surveillance Counter Measurers*, DOI 10.1007/978-3-319-28666-2_3

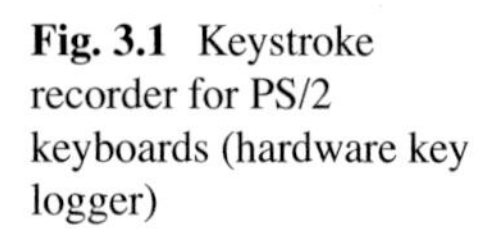

Fig. 3.1 Keystroke recorder for PS/2 keyboards (hardware key logger)

keystrokes are also recorded. If the perpetrator has more time at his disposal, he may opt for a more stealth option, that is, to install the recorder inside the keyboard itself or replace the keyboard with an identical one that has a logger already installed.

3.3 Means of Data Transfer

As we have previously mentioned, there are dozens of ways to transfer data. These ways include floppy disks, USB drives, CD/DVDs, cellphones, MP3/4 players, as well as using network services (email, FTP, VPN access, etc.). Let's only remind the leakage of confidential documents in the WikiLeaks case. Soldier Bradley Manning revealed that he has been copying classified data on rewritable CDs, going through security checks by stating that they contained music by Lady Gaga!

On the same wavelength, we should not forget that all the modern means of digital entertainment and mobile phones as well have storage capabilities of many gigabytes. Thus, digital cameras, cellphones, gaming consoles, printers, wristwatches, and dozens of other gadgets can "accommodate" in their internal memory thousands of company secrets.

Apart from digital information, a spy may possibly intercept data already printed on paper. Modern digital cameras have such a resolution that can clearly "photocopy" any document very easily by just taking pictures. More impressive tools include micro-scanners such as the one shown in Fig. 3.2. At the size of a marker, it scans and stores in its memory many pages of confidential documents. A very confident criminal with excessive audacity may even print the documents using the victim's printer!

More exotic means of hardware interception include supersonic transmission of information using the computer's loudspeakers, as well as hidden microsize Wi-Fi adapters that leak information via wireless connections.

Fig. 3.2 Portable marker-sized digital scanner

3.4 Software

Proceeding on to the software, things get even more challenging, since physical evidence no longer exists and the discovery of the threats can only be done using forensic analysis. A plethora of malicious software, viruses, and Trojan horses lurk to intercept our data.

Similar to the key loggers that we previously mentioned, there is also software that can serve the same purpose. In fact, such software is able to take a screenshot in regular intervals and relay the activity taking place on the compromised computer in real time. Indeed, a very effective way to fight against hardware-based key loggers is by using "virtual" keyboards (on-screen keyboard) where the user types with the help of the mouse, clicking on the characters he wishes to type. Obviously, since keys on the keyboard are not used, the key logger cannot record anything. However, the software-based version of the key logger has the intelligence to take a screenshot on every click of the mouse, thus easily revealing a password. Figure 3.3 shows a screenshot where a virtual keyboard is used. The malicious software was able to identify that the key "3" was selected. More advanced versions of such software are capable of activating the microphone that may be present or even the camera, turning the computer itself into an excellent "bug."

Other forms of malicious software allow the full and remote access to the victim's computer without ever being noticeable to the average user. The attacker has the capability to browse through the contents of the hard drive and easily copy anything he wants, as if he were sitting on the computer himself. Besides, if the legitimate user has no relevant training, it is possible for the attacker to use remote access software that does not even hide its presence at all. Many such software versions are by default installed in operating systems to provide remote assistance.

If the computer is password protected, then the attacker will try to boot the computer from a CD or a USB stick by using small-sized operating system that do not require a hard drive to boot since they run entirely on the removable media (e.g., live CD). That way some of the security measures are bypassed and access to the hard drive is given. In case the hard drive is encrypted, if the criminal has enough time at his disposal, he may try to "clone" the hard disk so he can later recover the data using cryptanalysis software at his premises. Usually the spy aims to cover his tracks to be able to have access again in the future. If that is not required, then he can simply just remove the hard disk drive and take it with him. It is no coincidence that there are drive bays that can be locked!

Fig. 3.3 Virtual keyboard with the user's keystroke tracked as recorded by keylogging software

Unfortunately, there are certainly even cases where a computer is left completely unprotected, even without a logon password being set (or possibly the attacker has already extracted the password in another way). Copying data is of course nothing unusual, unless the administrator has revoked access to the computer's removable media (e.g., USB, CD). For that kind or protection to be effective, the possibility of using email or a file transfer service in the Internet to send data should be taken into account. For these reasons, using a DLP (data loss prevention) system is imperative. This kind of software identifies what information is classified and does not allow the transfer of it outside the corporate network in any possible way.

But again, the spy will try to copy the data by taking a picture of the computer screen itself! Indeed, in order to locate the files that contain the desirable information, he may use specialized software that search the hard drive for specific type of files or keywords with different criteria. That kind of software is much more effective and faster than the simple "Search" that operating systems include. They are most commonly used by law enforcement for digital forensics. In addition to that he will also search for deleted files that could potentially contain classified data. As is already well known, when deleting a file, the data that it contained remain on the hard disk until the area they occupied gets overwritten by another file. Of course, the attacker again needs to have plenty of time at his disposal to conduct that analysis. A relevant forensics technique is that of file carving, that is, the process of reassembling computer files from fragments.

Software does not run solely on computers. All modern electronic devices are based on microprocessors or microcontrollers that run programs like computers do. For example, the Stuxnet worm was based on attacks against industrial control systems and more specifically programmable logic controllers (PLCs). Security regarding software for embedded systems is definitely a field that will concern us in the years to come, especially given the advances in the world of the Internet of Things (IoT).

3.5 Wireless Networks

Wireless networks are widely adopted in corporate environments. Unfortunately though, they inherit many of the security vulnerabilities and flaws of traditional wired networks, and additionally they are vulnerable to new threats. Radio waves travel freely without being easily confined; thus, an attacker may intercept data without ever being in close proximity to his target. Using directional antennas and the appropriate equipment, it is possible to conduct interceptions from great distances.

Additionally, communications maybe be hampered or even disrupted from denial of service (DoS) attacks caused by interference. As far as privacy is concerned, it is possible to record a user's position (especially in the case of cellphones). Particularly dangerous is the practice of connecting to uncertified and rogue wireless access points since they can be controlled by an attacker. This is very often the case in airports or other public places.

Encrypting the wireless link, on a higher level with encryption protocols as we previously discussed, further enhances confidentiality. Changing the default SSID name and all default passwords as well as disabling wireless management adds another level of security. For securely using wireless networks, it is imperative to enable encryption and more specifically the WPA2 encryption algorithm instead of the weak WEP one. A more thorough analysis on Wi-Fi encryption algorithms follows in the next paragraphs.

Regarding Wi-Fi networks, there are two main protocols used for secure communications: WEP and WPA/WPA2. The WEP protocol uses the RC4 (River Cipher 4) algorithm for encryption which suffers from multiple vulnerabilities and is now considered insecure. The RC4 algorithm is a stream cipher for symmetric encryption which produces a pseudorandom bit stream which combined with the encryption key that produces the cipher text after passing it through a XOR operation. To produce the cipher text, 24 bits of the initialization vector (IV, a fixed-size input) are used along with the pre-shared encryption key selected by the user which has a length of 40 or 104 bits. The result is passed through a XOR operation with the original plain text to produce the cipher text.

The way the WEP protocol combined the IV with the encryption key to initialize the pseudorandom bit stream (known as seeding) required high entropy for selecting good seed data. Instead of that, the length of the IV that was used was only 24 bits (small number of keys could be generated), leading to reused IVs with the same key (low entropy). It is worth noting that it was not the RC4 algorithm itself that it was vulnerable but the way it was implemented in the WEP protocol. The larger the number of packets the access point sent, the more possible it was for reused IVs with the same key to be sent with another packet. So, after capturing enough packets, it was possible to crack the cipher text and discover the encryption key. To speed that process up, there are techniques used to force the access point to resend packets because of packet loss, packet rejection, or sending NACK responses. That way the access point is retransmitting constantly and it is just a matter of minutes to send a pack with a reused IV.

There were various attempts to fix those issues such as by making the IV longer but that only made the attack take longer time. With the simultaneous advance in the processing power of CPUs, it didn't make any significant difference so the protocol remained insecure.

The WPA protocol also used the RC4 encryption algorithm with a 48-bit IV and a 108-bit user-defined encryption key. The most important security aspect that the WPA protocol introduced was the use of the Temporal Key Integrity Protocol

(TKIP) which issued a per-packet key as in dynamically generating a new 128-bit key for each packet. That way the WPA protocol was both compatible with older WEP equipment as it was based on the RC4 algorithm and immune to WEP-based attacks. The WPA protocol also included a message integrity check algorithm called Michael, a part of TKIP, which was much stronger than CRC used in WEP that was part of the vulnerability that could be exploited as we described above. Later it was discovered that a flaw in WPA that relied on older weaknesses of WEP and the limitations of Michael could be exploited to retrieve the keystream from short packets to use for reinjection and spoofing. The WPA protocol was used as an intermediate measure to replace WEP and until the WPA2 was widely available. Today WPA is considered to be insecure.

The WPA2 protocol finally replaced the RC4 algorithm with the AES and CCMP. Like TKIP, CCMP uses a 48-bit IV with a 128-bit encryption key. Instead of a sequence counter per packet, it uses AES keys to protect the integrity and confidentiality of the packet. There are various attacks known for WPA2 but it is considered relatively secure. One of the most common and interesting attacks is the one regarding WPS-enabled routers and access point. At this point we need to mention that this attack does not rely on the WPA2 protocol itself but on the implementation of WPS with WPA2 and other protocols. The attack can be deployed against the WPS PIN authentication system which consists of an 8-digit PIN. Since the last digit is a checksum of the previous digits, there are only seven digits left to guess. The WPS system validates the PIN in two halves and that's where the vulnerability is found. The first part of the PIN consists of four digits, while the second part has three digits. Since they are validated separately, an attacker could also guess the two parts separately, guessing the second part after guessing the first part successfully. This means an attacker only needs 11,000 guesses (10,000 for the 4 digits and 1000 for the 3 digits) to gain access to the network which can be done in under 4 h. Since this attack is based on the way WPS is implemented, there were occasions that even disabling the WPS function didn't make the network immune to this attack.

Updating the drivers and the firmware of the device is again imperative, while more advanced methods to add security include access control lists based on MAC addresses and using static IP addresses instead of dynamic ones through DHCP. Apart from lowering the transmit power, placing the access point in the middle of the office instead of near the windows can also reduce the radio wave footprint.

Any type of wireless network implies such a great risk that the US National Institute of Standards and Technology (NIST) recommends all agencies to not use wireless networks for their critical operations before ensuring the mitigation of threats deriving from their use on all levels, both regarding the information transmitted through wireless networks as well as regarding their business continuity (NIST SP 800-48, Guide to Securing Legacy IEEE 802.11 Wireless Networks, July 2008). Given that cellphones are wireless systems, this recommendation could be expanded to include cellphones as well.

3.6 Man-in-the-Middle Attacks

Man-in-the-middle attacks can be thought as simply intercepting a communication between two systems. This usually involves a system pretending to be a legitimate part of the communications channel, thus routing the information through it, without the original two systems being aware of that.

Using man-in-the-middle attacks, an attacker can alter the contents of the communication before forwarding it its original destination, cause denial of service (DoS), or simply eavesdrop on the communication.

There are many practices that can be considered as man-in-the-middle attacks. The most common one regarding mobile communications is by setting up a rogue base station, tricking the target cellphones in a certain area to connect to the rogue base station and then relay the communication to the legitimate base station as it is extensively described in another chapter.

That way, the attacker has full control over the communications that pass through the rogue station and can be recorded or altered. Regarding computer networks, the most common practice to deploy such an attack is ARP poisoning. In a similar manner to the mobile communications practice mentioned above, the attacker tricks computers into acknowledging the attacker's equipment as a legitimate part of the network they are connected to that needs to receive the traffic. Again, the attacker can choose to relay the communication to seamlessly eavesdrop the communication, alter its contents, or just reject it to perform a DoS attack.

Address resolution protocol (ARP) is a protocol used for the resolution of network layer addresses into link layer addresses and used in many protocol implementations such as Ethernet. That means such an attack is effective in both wired Ethernet networks and Wi-Fi networks. Within a network, when and IP packet is sent from one host to another, the destination IP address needs to be resolved into a MAC address for transmission in layer two. When the host's IP address is known and the MAC address is needed, an ARP request is broadcasted in the network so the host with that IP address with reply to announce its MAC address.

The ARP architecture suffers from two main problems. Network hosts cache ARP replies even when they weren't requested by another host as well as they will overwrite older cached entries with new ones when they are broadcasted even when the older ones have not yet expired. Furthermore, there is no mechanism in the ARP protocol to authenticate the origin of ARP replies. These two vulnerabilities allow for ARP poisoning attacks to be successful.

An attacker can associate his MAC address with the victim's IP address so any traffic that would originally be forwarded to the victim would then be forwarded to the attacker's host instead. As with other similar implementations of man-in-the-middle attacks, the attacker can choose to simply eavesdrop on the traffic while forwarding it to the original destination, altering its contents before forwarding it or dropping packets to cause a denial of service.

3.7 Data Security

Training needs always to be one of the first steps and concerns in every effort to secure information. It is imperative for users to be aware of the principles of data security as well as of the relative best practices. Specifically those individuals who handle sensitive data as well as those who transfer such data outside the organization/company ought to fully understand the value and sensitivity of that kind of information. Of course it is mandatory for the IT department to have adequate resources available, in order to maintain experienced and constantly trained technical personnel capable of properly supporting the users.

Before getting to the technical means, we need to establish the appropriate policy guidelines. Will the company provide laptops and smartphones to the users or not? Is it allowed to browse the Internet? Will teleworking be used? What kind of data is allowed to be stored on the computer and what isn't?

Following those questions, decisions regarding aspects relevant to policies and processes for handling any electronic means of storage, processing, and data transfer must be taken. These policies will, among other things, specify what software is necessary and certified to be used on the company's computer systems. Any other software that is allowed to be installed on computers is defined beforehand in order to ensure secure operations and interoperability with the rest of the systems.

In case an employee is using his own computer, it is not always possible to restrict him from using noncertified software, which inevitably leads to the need of taking technical measures to enforce the proper behavior. It is important to note that, regardless of restrictions, many users will still want to install instant messaging (IM) and peer-to-peer (P2P) applications. It is therefore imperative to implement and enforce administrative and technical mechanisms in order to assure the security compliance since there will always be employees that will, deliberately or involuntary, disregard such instructions.

Respectively, clearly communicated processes for security issues and incident reporting allow the employee to act immediately, thus limiting the impact that any particular incident may have. As a whole, adopting an information security management system (ISMS) such as ISO 27001 represents a good first step toward revealing risks as well as toward establishing security policies and procedures and defining the need of stuff training for responding to those risks.

Computers are obviously the single most important tool of modern businesses. However, software running on those computers may not be correctly configured or it may not be updated any more. It is also very likely to use an insecure wireless network or to be infected by malware from other computers on the network, especially if these computers access the Internet. Thus, it is imperative to use computers that are dedicated solely to business needs along with the appropriate security policy and documented processes. Strong passwords and anti-malware software should be combined with regular system updates and upgrades. The latter two may also be automated using a patch management system. Such systems monitor the computers of a network and automatically send the correct patches to be installed.

Administrators are generally able to remotely manage computers not only for the daily operation but also to conduct security checks and audits. In case that is not possible, either because of policies or because of lack of resources, users may take part of this responsibility themselves. Again, appropriate training plays a decisive role. Other tools can monitor the network's traffic and it general behavior to spot possibly dangerous deviations that are signs of attack or leak. Network access control software proves to be a valuable tool since it checks whether the operating system and software of a particular computer is properly updated which decides whether it will be allowed for it to be connected to the corporate network or not. Finally, software for deactivating the USB ports can dramatically limit the easiness of data and trade secrets being leaked outside the company.

Especially for laptop computers, it would be a good practice not having distinct characteristics that may reveal that they are corporate laptops like logos and stickers, since this way they can be targeted much more easily. Since laptop theft is one of the most common ways of data leakage anyway, it is only logical trying to disguise them as personal computers.

Nevertheless, if someone had to choose a single one security measure, then it shouldn't be any other than encryption. With the appropriate encryption framework, even with the lack of other security measures, information is sufficiently protected (the question in this case is for how long it will be protected, since developments in cryptanalysis can render a previously thought to be secure algorithm weak).

Besides communication encryption, stored data should also be encrypted, without forgetting encrypting backups. That way, even in the case of a security breach, or even in case of stealing the computer, extraction of data will be particularly hard if not impossible.

3.8 Data Traffic and Management

Limiting the data being transferred around personal computers and laptops is one basic measure of information security. It is preferred for data to be protected on a server and not being dispersed on computers and laptops. Access to that data will only be allowed for authorized users, while downloading, storing, and sending it via email or any other way will be forbidden. In that respect, cloud computing services seem to be heading in the right direction as long as the service provider can actually guarantee a proper security level.

Of course, there are cases where data transfer is needed (e.g., during transportation of backups). It is well known that backup data is required to be in a different location than the original computers that it stemmed from. Therefore, since transferring data is necessary, encrypting them gets mandatory. The removal of data in this case should be done in a strictly controlled manner. Obviously we should not forget to encrypt any removable storage media such as external hard drives of USB sticks.

Data should be partitioned into fragments with employees having access only into these fragments that are needed to complete a given task. They must be available only for the duration of the given project/task, and access should be immediately revoked after completing the task. Another good practice is for that data to be stored on temporary-intermediate servers. In that way, even with a successful attack or interception of data, the impact will be confined only to that specific data subset.

Full logging and monitoring of access to the data is also necessary. Among other things, logging includes the time, date, username, workstation, files, and records being accessed, and it can be an important aspect of an incident handling procedure. Data loss prevention (DLP) technologies and data watermarking (for data leakage tracing) can prove to be very powerful tools in that direction.

Regarding communications they should only take place easily encrypted channels. The usage of virtual private networks (VPNs) is the most common practice. They route private traffic through encrypted tunnels inside the generally untrusted provider's network while offering security, without the need to invest on a dedicated network or expensive leased lines.

Last but not least, user authentication shouldn't rely solely on a password. Methods of multiple authentication layers include smart cards and one-time-password (OTP) devices.

3.9 Advanced Persistent Threats

The advanced persistent threat (APT) term describes the processes and practices involved in order to get access to data or achieve a specific mission against a given target. They are characterized by the very specific target against which they are deployed, their covert operation, and the time they are active since usually they are deployed for lengthy periods of time and operate in a continuous manner.

Since the development of APTs requires a high degree of expertise, they are used where the gain is also significant whether that gain is financial or in a manner of information-related advantage. APTs are very often deployed against large organizations and governments as their nature makes them one of the best tools for conducting espionage, industrial or not.

Having a very specific target means not only they are custom made for each organization or government they will be deployed against, but usually they target a single system and they have just one or small number of objectives. Their persistent nature means they usually communicate with a command and control server (C&C) and continuously intercept information until their goal is met.

The most common case that can be considered as an APT is the Stuxnet worm that infected one of Iran's nuclear plants. As with most APTs, Stuxnet had only one goal and targeted one system: the system (PLCs) that controlled nuclear centrifuges in order to spin them out of control and destroy them. Even though Stuxnet infected a large number of computer resources, it didn't do any damage since it only did that to gain access to the target system.

The nature of APTs makes them very hard to detect, so continuous monitoring with various security tools and appropriate information security personnel training are the best strategy to mitigate threats like that.

3.10 Teleworking and Industrial Espionage

The assumption that the work environment consists solely of the employee's office and that interception may only take place there is obsolete. Thanks to teleworking, a flexible form of work which allows an employee not to be present at the employer's premises but to be working from a remote location, security risks are increased. In the most common form, that location will be the employee's home. Taking into consideration employees such as sales staff who are traditionally outside the company's premises for long periods of time, then teleworking can be taking place at a hotel or even on the go, for example, on board a train. The implementation of teleworking is possible by using technical methods and services that involve both IT and network assets. From the IT perspective there are tools such as laptops, faxes, and smartphones. Respectively, from the network perspective, the communications channel can be a DSL line, a wireless network, a cellular network, or even a satellite (VSAT).

Despite the broad acceptance of teleworking, several studies indicate that there are always some concerns that slow down the even wider adoption of teleworking. Some of those concerns focus on management challenges, disruption of workflow, and loss of contact with clients and colleagues. The most important factor though is none other than security concerns and consequently possible abuse in cases of industrial espionage.

Indeed, in one of the biggest studies on that matter (2008 CDW-G Telework Report), concerns regarding information security is the main reason why the adoption of teleworking is delayed. As shown in Fig. 3.4, this reason accounts for 40 % of the public sector and for 27 % of the private sector, being the most important reason for both sectors.

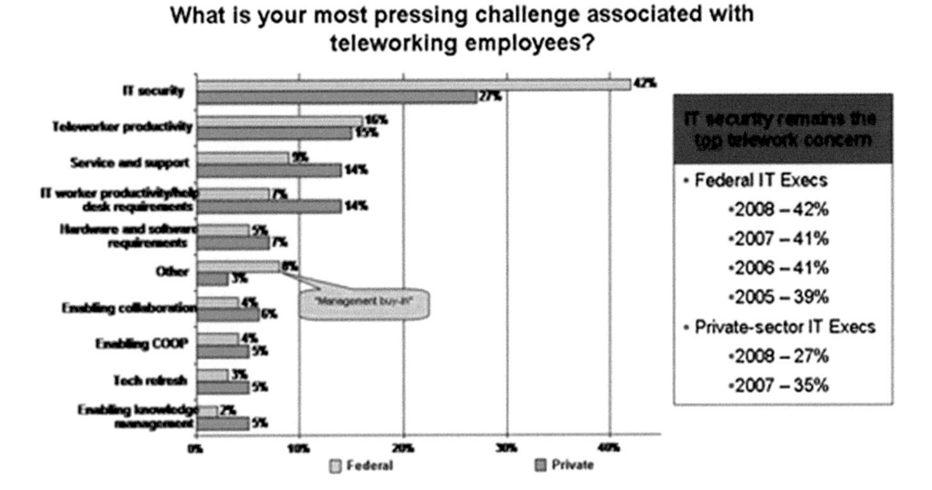

Fig. 3.4 Factors for delaying the adoption of teleworking (Source: 2008 CDW-G Telework Report)

At first glance, the concerns that executives have are justified since there is a number of threats that affect the availability as well as the confidentiality and integrity of data. The combination of different locations and different communication channels and a plethora of tools pose challenges for security that need to be addressed both individually but also as a whole within the teleworking framework.

Among these challenges, physical loss (or theft) of portable devices is dominant. Laptops, smartphones, external hard drives, and USB sticks can become an easy target due to their small size. Besides, different interception devices or malware can have unpleasant consequences. Plain shoulder surfing taking place in a public place can also disclose confidential information! Let's don't forget that at home, computers may be used by other members of the family, making the problem even more acute, since the employer has less control. Also, home computers are usually less protected than the ones in an office and generally are less controlled.

Apart from malicious acts, a simple hardware failure could disrupt the communication of the user with the company without an immediate solution being possible, which would otherwise be possible to offer at the company's premises. Respectively, the practice of outsourcing the helpdesk and remote equipment maintenance services paves the way for further risks concerning industrial espionage.

So, what can go wrong? From the dozens of documented incidents over the past few years, we cite some with the highest impact:

- On May 2006, a laptop containing personal data of 26.5 million US Army veterans was stolen. As it was proven later, the employee was transporting the laptop to his house without prior authorization and without the necessary security precautions.
- Respectively, on November 2007 there was a loss of two CDs containing personal data of 25 million citizens that were being sent by plain mail service (and without being encrypted) from Her Majesty's Revenue and Customs in the UK.
- Again in Europe but in Germany this time, on October 2007, T-Mobile announced the loss of a disk containing personal data of 17 million customers.
- In 2007 a laptop of a VeriSign employee was stolen from his car. It contained personal information of an unknown number of VeriSign employees including names, Social Security numbers, dates of birth, salary information, telephone numbers, and home addresses. The data on the laptop was not encrypted in contravention of VeriSign policies.
- In a similar manner, in 2008, a laptop belonging to a Royal Navy officer was stolen from his car containing personal information regarding 600,000 applicants and people interested in joining the Royal Navy, Royal Marines, and the RAF. The data contained details such as passport numbers, National Insurance numbers, and bank account details.
- In 2008 HSBC lost an entire server during renovations in their offices in Hong Kong. HSBC stated that the chance of data being misused was very low, but since there was physical access to the server, attacks would be significantly easier. The server contained financial information regarding 159,000 customer accounts.

The "optimistic" side of a laptop theft is that the thieves usually do not realize the value of such data. They only care about the immediate reselling of the laptop. Obviously that does not apply to targeted thefts conducted as part of industrial espionage.

3.11 Global Surveillance

Even though over the years a lot of theories have been expressed regarding global surveillance programs, probably the most famous one was project PRISM. This was revealed in 2013 by Edward Snowden when classified documents about the program were leaked to journalists of the Washington Post and the Guardian.

The PRISM program involved the collection of stored Internet communications based on demands made to Internet companies. Through these demands NSA could use PRISM to target communications that they were encrypted when they were transmitted. The leaked documents identified several technology companies as participants in the PRISM program, including Microsoft, Yahoo, Google, Facebook, Paltalk, YouTube, AOL, Skype, and Apple even though there were various responds from these companies regarding their participation. Even though the PRISM program operated in the USA, it had a global effect since services of these companies are offered to users across the globe.

A much less known mass surveillance program which was also revealed by Snowden was MUSCULAR. Jointly operated by Britain's GCHQ and NSA, it involved secretly tapping the main communications links connecting datacenters of Yahoo and Google.

The MUSCULAR program operated via an access point outside the USA and the exploitation was based on the fact that data inside Google's private cloud was being transmitted unencrypted (possibly to avoid the encryption's overhead) with Google's frontend servers stripping and, respectively, adding back SSL from and to external connections. According to the leaked documents, millions of records of data intercepted every day using the MUSCULAR program were sent to data warehouses at Fort Meade.

After the disclosure of the program's operations, Google announced that it was encrypting traffic between its datacenters and Yahoo followed with a similar announcement.

3.12 Summary

In this chapter we have examined the software and hardware being used for conducting industrial espionage, when computers and data being processed by them are targeted. As mentioned in the introduction, the full description of the techniques and the threats that may compromise the confidentiality of data are not possible to be documented in the few pages of a chapter since it is the subject of the information security science. The interested reader will easily find further information in the hundreds of relevant book titles as well as in the thousands of websites on the Internet. Continuing onto the next two chapters, we will be examining quite more extensively voice and data interceptions in both landline and mobile communications.

Chapter 4
Intercepting Fixed Line Telephony

4.1 Introduction

The first thing that comes in mind when hearing the term "eavesdropping" is probably intercepting telecommunications and, more specifically, telephone conversations. In the case of industrial espionage, eavesdropping on the conversation between executives can be of more value than the technical information contained in thousands of pages of documents that are intercepted from computers. Indeed, such a conversation could reveal the next business strategy, pricing policy, and information regarding bids for tenders, mergers and acquisition details, introduction of new products, and so forth. In this chapter we will describe interceptions conducted on fixed line telephony. Interception devices may be placed on the telephones we use or on the cabling and the network and provider's equipment. Both of hardware means but also software means do exist. Respectively, the usage of newer technologies for fixed line telephony, such as VoIP, "inherits" security threats from the Internet and implementing those requires a certain level of awareness.

4.2 Fixed Line Telephony Interception Devices

Telephone conversations can be intercepted in real time or recorded in order to be retrieved and analyzed later. Calls can be eavesdropped from internal or external monitoring points or even from monitoring stations located in other countries. Let's not forget that modern telecommunication networks handle data as well, such as fax and Internet data. So, with the appropriate equipment data can also be intercepted along with voice conversations.

Especially concerning fixed line telephony, the presence of voltage on the wire itself allows the continuous and long-term operation of a transmitter bug without the need for replacing batteries such as Fig. 4.1 depicts. The latter can even be placed

I. Androulidakis, F.-E. Kioupakis, *Industrial Espionage and Technical Surveillance Counter Measurers*, DOI 10.1007/978-3-319-28666-2_4

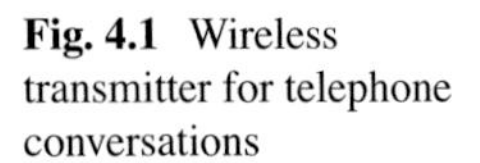
Fig. 4.1 Wireless transmitter for telephone conversations

inside the telephone device itself. The picture of a movie "Spy" who unscrews the handset's microphone on an old analog telephone device is very typical even though the technology used today is different.

Another basic element that is worth mentioning is that of a device installed in a telephone device that can also intercept ambient conversations taking place inside the office and not solely during calls. This device can be "woken up" and start relaying audio from the microphone of the phone, without the phone ever ringing.

4.3 Wiretaps

Other traditional methods such as clip-on, where a wire pair is connected in parallel with the original one, are equally effective. Wiretaps that either record or transmit the voice are more common. In this case, the device can be connected to any point across the length of the line, starting from the main distribution frame (MDF) at the central office (CO), passing through the local distribution cabinets (also known as KVs from the German word "Kabelverzweiger"), continuing onto the demarcation point (where the public-switched telephone network ends and connects with the customer's on-premises wiring), reaching the building's distribution frame, then onto the vertical cabling, and finally even in the telephone device itself.

Figure 4.2 shows a KV while Fig. 4.3 shows a distribution frame (Fig. 4.3 is a magnified view of Fig. 4.4). Readers can notice a common maintenance practice that always makes interceptions easier. Sensitive information about the local loop, the interconnections, and the cable routings are noted on pieces of paper left by technicians during installation to ease the process during future visits. Even worse, we can notice a tool for parallel connections which taps onto the subscriber's line termination slot and provides access to the line without breaking the circuit. Even with digital telephone exchanges where the switching on the backbone is done digitally, subscribers' telephone devices on the access network can be analog, so the traditional methods that were mentioned above are still effective (Table 4.1).

Fig. 4.2 Local distribution cabinet-KV

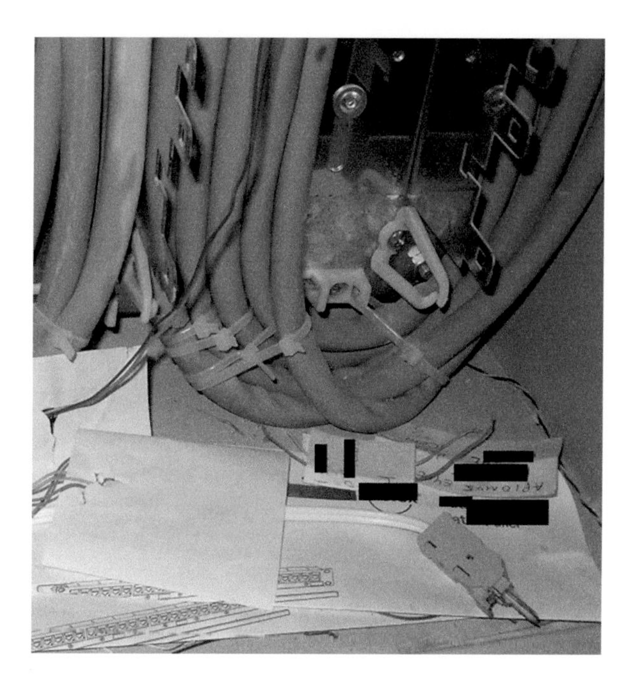

Fig. 4.3 Sensitive information regarding the interconnection of the lines and the tapping tool left at a distribution frame

Fig. 4.4 Medium-sized distribution frame

Table 4.1 Nontechnical means of security

Nontechnical means of security
Apply formal procedures and security policies
Train the users, the administrators, and the operators/agents
Document the entire infrastructure (equipment, cabling, interconnects, configurations, etc.)
Conduct regular visual inspections and audits
Include telephony equipment in the business continuity plans
Make regular backups and check their integrity
Consider insuring the equipment by insurance companies
Make mutual deals with other companies in the same building to "lend" telecommunication services during an emergency
Enforce processes for incident management
Keep regular contact with the equipment suppliers and the provider
Stay up to date with the current threats and vulnerabilities in any possible way
Use document shredders
Do not post internal phone directories online
Be cautious with shared telephone devices (e.g., conference rooms)
Disconnect the telephone devices during confidential meetings
Apply an information security management standard such as ISO 27001

4.4 Structured Cabling

In order to interconnect hundreds of telephone devices within a company, a structured cabling network is necessary. Distribution frames across the length of that network are responsible for distributing the signal, and along with the transmission lines, they make up the backbone of the telecommunications' infrastructure. The main distribution frame is the point where internal and external lines are interconnected. In the case of a private branch exchange, one part of the distribution frame connects to the private telephony exchange (PBX)'s line cards, and the other part connects with the users and the external network. Figures 4.5 and 4.6 illustrate main distribution frames of various sizes. The access cabling that reaches the users may pass through intermediate distribution frames (e.g., on every floor) and wiring closet racks such as the one shown in Fig. 4.7.

The cable routing of the internal network can span hundreds of meters, passing through different ducts, false ceilings and floors, distribution frames, and patch panels. So, it is possible to install an interception device or perform a clip-on, anywhere across the length of that cabling infrastructure. Indeed, as shown in Fig. 4.8, it can prove very challenging to detect a suspicious wire pair among hundreds of other pairs of the same color located at the distribution frame (Table 4.2).

Fig. 4.5 Small office distribution frame

Fig. 4.6 Large distribution frame (>1000 connections)

4.5 Digital Lines and Private Branch Exchanges

In analog telephony, voice is transmitted as analog current. Even by connecting a plain loudspeaker parallel to the line one can pick up the audio. Regarding purely digital lines (such as ISDN, private branch exchanges with the appropriate digital telephone devices, as well as VoIP), sound is transmitted in the form of digital data packets. In that manner, it is not possible to simply tap the line with a secondary device or with an analog interception setup. Unfortunately though, if no encryption techniques are used, those packets can be assembled back and converted back to audio and reveal the content of a conversation. Especially regarding private branch exchanges (PBXs), there are specific devices, depending on the brand and type of the exchange, that do exactly that, convert digital signal into analog sound. Respectively, for VoIP, things are even easier since there is a software that can do the same thing.

Fig. 4.7 Wiring rack

Besides, in the case of private branch exchanges, telephone devices can be modified (or programmed) to transmit ambient audio even when they are inactive (on hook). In that way, they allow eavesdropping on conversations taking place at the office they are located in. So, imagine the telephone set in a company's meeting room transmitting all the company's critical conversations to another person. The important point here is that it is impossible for a cheap analog device to operate in that way with programming. It would require physical intervention. On the other hand, devices of hundreds of euros, with digital displays and dozens of functions and buttons, can be programmed or "ordered" from the exchange to operate the microphone and make a call remaining completely silent and with no indicators switched on whatsoever. That way it can be transformed in an excellent "bug."

Fig. 4.8 Magnified part of Fig. 4.6

Table 4.2 Technical means of security

Change passwords regularly on accounts and devices
Use strong authentication of two or more layers
Deactivate remote management and administration on the telephone exchange
Separate management network from the production network
Deactivate nonused functions such as dial through or DISA
Use lengthy PIN codes
Use the "#" symbol to terminate PIN codes
Check and audit the exchange's configuration regularly
Replace the analog cordless telephones with digital ones
Check the security of DECT devices with tools available online
Consider having communications encryption using crypto phones

Respectively, the telephone exchange (with its microprocessor and the software that controls it) is possible to be wiretapped using the appropriate software. That software can be part of the exchange's diagnostics or part of lawful interception functions. Lawful interception refers to obtaining communications' network data pursuant to lawful authority for the purpose of analysis or evidence. For example, a specific type of exchange (with the ability to handle thousands of users) uses the command "listen." It is that simple for the administrator to request eavesdropping between the victim's telephone set and any other set on the exchange. As such, the CEO's telephone set can be wiretapped so the conversations can be rerouted to another device of an employee who can record them by using a digital recorder.

Moreover, it is possible to transfer the intercepted call to an external line so the attacker doesn't even need to be present at the location of the interception. Except for the commands available natively in modern exchanges, third-party software can also be used, especially in solutions based on open-source systems such as Asterisk VoIP.

At the same time, system administrators can activate line supervision and monitoring functions. Abusing these functions allows eavesdropping on conversations taking place on those specific telecommunication systems. They are using the same functionality that allows legitimate applications on the exchange such as the call monitoring of operators/agents from their supervisor for performance and quality control, intrusion into calls, as well as three-way calling services. In all cases of abuse, the third member of the three-way call is of course the eavesdropper.

Leaving the malicious equipment aside, there is also the matter of lawful interception that telecom providers are obliged to practice when requested by the authorities. While procedurally it is required for a prosecutor's order to be issued, technically things are much simpler. So, lawful interception systems abuse, with or without the provider's cooperation, can easily be exploited against targets for industrial espionage purposes.

Usually, when legitimate eavesdropping intrusion tools are used, by using commands of the exchange, as well as during three-party calls, there is a warning tone beeping at regular intervals (e.g., every 15 s). That tone informs and reminds the conversing parties that there is a third party involved and that the call might be recorded. Readers may have noticed that warning tone when calling a bank to make transactions by phone. Unfortunately, very few are aware of the importance of that warning tone due to the lack of information. What is more worrying is that administrators can deactivate the tone altogether. Even if the system does not allow for the deactivation of this warning tone, it might allow other configurations that effectively make the tone disappear. Indeed, it can be possible to set the duration of the tone to a fraction of a second or set its frequency very high so it would be in both cases impossible to actually be heard from the telephone's handset.

4.6 Call Detail Records Interception

During industrial espionage, it is not always necessary to intercept the content of the conversation itself. In many cases it is only enough to know the contacts and parties that a company is communicating with. Intercepting call detail records (CDRs) can provide valuable information. Call detail records keep information about calls, such as the calling and the called party, the duration of the call, the trunks that were used, and so on. An example can be found in Fig. 4.9.

Correlating the calls with special telecom traffic correlation techniques, interesting connections, private contacts, and professional partnerships (e.g., with suppliers) can be revealed. For example, if the call detail records show a significant number of calls being placed between the company and the patent office as well as between the technical and legal departments of the company, it is a safe assumption that a new

```
----[/DHS3dyn/account/TAXATGHP.DAT : Ticket number ■■■■]--------------------
<00>              TicketVersion = ED5.1
<01>               CalledNumber = ■■■■
<02>              ChargedNumber = ■■■■
<03>            ChargedUserName = ■■■■
<04>          ChargedCostCenter = ■■■■
<05>             ChargedCompany =
<06>           ChargedPartyNode = 105
<07>                 Subaddress =
<08>              CallingNumber =
<09>                   CallType = PublicNetworkIncomingCallToPrivateNetwork
<10>                   CostType = ISDNCircuitSwitchedCall
<11>                EndDateTime = 20080826 01:35:40
<12>                ChargeUnits = 0
<13>                   CostInfo = 0
<14>                   Duration = 417
<15>              TrunkIdentity = 360
<16>         TrunkGroupIdentity = 1
<17>                  TrunkNode = 101
<18>         PersonalOrBusiness = Normal
<19>                 AccessCode =
<20>         SpecificChargeInfo =
<21>           BearerCapability = Speech
<22>              HighLevelComp = Unspecified
<23>                 DataVolume = 0
<24>           UserToUserVolume = 0
<25>           ExternFacilities =
<26>           InternFacilities = BasicCall
<27>              CallReference = 0
<28>              SegmentsRate1 = 0
<29>              SegmentsRate2 = 0
<30>              SegmentsRate3 = 0
<31>                    ComType = Voice
<32>        X25IncomingFlowRate = Unspecified
<33>        X25OutgoingFlowRate = Unspecified
<34>                    Carrier = 0
<35>        InitialDialledNumber = ■■■■
<36>            WaitingDuration = 1
<37>       EffectiveCallDuration = 417
<38>     RedirectedCallIndicator = 0
<39>              StartDateTime = 20080826 01:28:43
<40>       ActingExtensionNumber =
<41>           CalledNumberNode = 9999
<42>          CallingNumberNode = 9999
<43>   InitialDialledNumberNode = 9999
<44> ActingExtensionNumberNode = 9999
<45> TransitTrunkGroupIdentity = 32767
<46>              NodeTimeOffset = 0
```

Fig. 4.9 A typical call detail record

product is about to be patented. In the same spirit, intercepting telephone numbers located on the telephone device's memory or on the call forwarding tables can lead to exposure of the private telephone numbers of important executives. Respectively, stored passwords or passwords used for telephone banking may be leaked as well.

4.7 Wireless Communications

The interception techniques that we have examined so far are performed in an active manner. Thus, there is some kind of physical or technical interaction with the telephone device or the provider's network that causes changes and leaves traces. These changes constitute a direct indication that the system's confidentiality and integrity have been compromised. Physically placing a wiretap is a typical example. Correspondingly, the reprogramming of a telephone exchange involves a technical interaction whose results can be traced, both on the system's audit files and from the changes in functionality. In a few words, active interceptions can potentially be revealed. How easy or hard this can be done is of course altogether another matter. With the help of this book we hope to make this task easier thus minimizing the attack impact.

On the other hand, there are also passive interceptions. When such an interception is performed, there isn't any kind of interaction, just passive "listening." A typical example is the one of directional microphones mentioned in the previous chapter. Regarding telecommunications, wireless systems are the "heaven" of every spy. Indeed, by simply "listening" to (sniffing is also used as a term) the radio waves that travel freely in the air, sensitive information can leak without anyone being able to notice it or even do something about it. Indeed, in a typical wireless communications environment, radio wave transmission is hard to confine; thus, potential eavesdroppers can capture and process signals without even being close to the "victim's" premises. Even encrypted communications can be intercepted and analyzed. If the encryption algorithms are not strong enough then with the appropriate demultiplexing and cryptanalysis techniques that target vulnerabilities in protocols and algorithms, the original content of the call can be captured.

More specifically, as far as telephony is concerned, the old analog cordless telephone devices (usually white colored) were nothing more than analog transmitters at 46–49 MHz band. With a simple radio frequency receiver, anyone could intercept conversations since there wasn't any kind of encryption used. Actually, with the appropriate antenna it was possible to eavesdrop even from more than 1 km away. Fortunately, the advent of DECT digital cordless devices solved the problem by offering digital transmission and encryption. However, like every other technology, DECT wasn't left unscathed. There are tools and software (even open source) for intercepting DECT transmission even though they require special hardware and their success depends on certain conditions.

Finally, let's not forget microwave links which connect remote offices or even the backbone of the cellular networks themselves. Even though the equipment required for intercepting them is notably expensive and requires, among others, visual contact within a very narrow field of view (due to their highly directional signal), intercepting them is still a possibility. The eavesdropper needs to simply be within the path of the microwave signal lobe. Depending on the distance from the transmitter, the useful beam width of the main lobe can be several dozens of meters, making the whole process easier. Often, these links aren't even encrypted. A very interesting example relevant to this can be found in "The Capenhurst Tower" case.

According to "Wikipedia," in 1999 the journalist Duncan Campbell published claims that a 50 m high tower on the premises of the uranium enrichment plant in Capenhurst had been used to intercept telephone calls transmitted by microwave between the British Telecom towers at Gwaenysgor, Clwyd, and Pale Heights, near Chester. Campbell claimed that the interception was conducted by the Government Communications Headquarters, GCHQ, initially from a temporary installation on the roof of the plant until commissioning of the tower in 1990. The main route for phone calls between Ireland and the UK was via the submarine fiber-optic cable UK-Ireland 1, landed at Holyhead, Anglesey, and then transmitted by a microwave link. Campbell claimed that calls were monitored by GCHQ until 1998 when the Irish telecommunication system was changed.

4.8 VoIP Specifics

Going back to the office surroundings, modern communication technologies face their own problems. They may have progressed and solved some vulnerabilities and older security concerns but opened the door for new ones. Telephony over the Internet by using VoIP or other technologies has become quite popular and is a typical example. Commercially, among other things, it is marketed as a more secure way of communicating thanks to the encryption that can be implemented. In reality though, that doesn't always happen, and it is often the case that the new system is less secure than the older classical telephony one. In this instance, the interception is based on monitoring, capturing, and reassembling the "voice" data packets that are transmitted within the network. If the communication is not encrypted, it is extremely easy to reconstruct the packets in order to come up with the original conversation. It is similarly easy to acquire the content from a video call. In a more advanced level, it is possible to reroute the packets by applying man-in-the-middle attacks as described in a previous chapter. These attacks exploit vulnerabilities in the protocol implementation and the protocols themselves, used for signaling in order to establish the call.

Another significant issue with VoIP communications is the lack of accurate authentication of the users. In traditional telephony, the phone number is bound to a physical address and a specific individual or a company. However, VoIP calls pass through the labyrinth of the Internet, and it isn't always easy to assess the physical determination of the call's location. A call may indeed originate from any computer that has internet access. In addition to that, caller ID can be an equally troubling situation. Like readers may have noticed, VoIP calls may not have a caller ID number or show numbers that don't correspond to a real number. Even worse, the process of deliberately changing the caller ID number, to display any desirable number, is relatively easy. This capability may be used along with social engineering techniques. A spy could extract information by making a VoIP call showing a number that is familiar to the victim. The spy then presents himself as a colleague, as a helpdesk technician, or even as a manager, deceiving the other party who has confirmed the identity of the caller solely based on the (fake) caller ID number (Table 4.3).

Table 4.3 VoIP security

Keep software on the servers and on the VoIP devices updated
Limit the unwanted services that run on these servers
Take action for the overall security of the IT equipment since VoIP relies on the current infrastructure (e.g., DNS, DHCP, etc.)
Separate the telecom traffic from the data traffic on the network
Activate TLS and SRTP

4.9 Telecommunications Security

Concluding the chapter, we will briefly describe some security measures regarding fixed line telephony and private branch exchanges. The full description of them is given in the book *PBX Security and Forensics: A Practical Approach* (SpringerBriefs in Electrical and Computer Engineering) (Table 4.4).

Table 4.4 Advice for end users

"Lock" the telephone set in your office when you leave
Regularly change passwords
Take note of the serial number of your telephone device
Personalize your telephone set so it cannot be swapped with another one without you noticing
Do not store passwords on the telephone's memory
Familiarize yourself with the device's indications and the exchange's warning tones (e.g., in case of three-party calling)
Do not rely solely on caller ID to identify the caller (as it can be spoofed)

Chapter 5
Mobile Phones Interception

5.1 Introduction

The universality of the usage of mobile phones and their penetration in everyday life are beyond any doubt. Modern mobile phones, smartphones, have advanced functions that converge with those of computers, and they are used not only for communicating but also for storing, organizing, processing, as well as sending and receiving data from their owner. Over time, their usage leads to the concentration of an increased amount of data concerning not only their owner but also the company he works for. That data may reside on the devices themselves as well as on the provider's computer systems. Therefore, dealing with the confidentiality of personal and business information is crucial. Especially in the framework of industrial espionage, besides intercepting voice calls, a cellphone may well be used not only as a bug but also as a means of unauthorized transfer of data and trade secrets outside of the enterprise environment. The latter can be achieved by using the phone's internal memory or the integrated camera. In this chapter, we will be examining all those matters in detail. We will start from the interception of data from mobile phones, we will continue onto the examination of network security issues, and we will conclude with techniques of intercepting conversations. The interested reader can find a more thorough description on the book *Mobile Phone Security and Forensics: A Practical Approach* (SpringerBriefs in Electrical and Computer Engineering).

5.2 Cellphone Data that Can Be Intercepted

The amount of data being stored on mobile phones is growing rapidly due to the more frequent usage of smartphones that replaced the previous generation of personal digital assistants (PDAs). Despite their unparalleled design, these handy smartphones introduce new threats. They contain a large volume of data (often of personal nature), but they can (easily) be stolen or lost due to their small

I. Androulidakis, F.–E. Kioupakis, *Industrial Espionage and Technical Surveillance Counter Measurers*, DOI 10.1007/978-3-319-28666-2_5

Table 5.1 Cellphone data that can be intercepted

Cellphone data
Contact list
Recent incoming/outgoing/missed calls
Incoming and outgoing SMSs and MMSs
Audio recordings/voice memos/ringtones
Pictures, videos, graphics, desktop elements
Calendars, alarms/reminders, to-do lists
Notes
E-mails
Browser history
Documents and files of any kind
User identification data (e.g., PIN)
Device identification data (e.g., IMEI)
Available networks list
Geolocation data (device location)

size. At the same time, the usage of smartphones can provide insight into their owners' personal and professional activity. Where does such data reside? Obviously, on the device itself but also on the network provider's computer systems. Especially regarding the device, data may reside in the internal memory, the SIM card's memory, as well as in the removable memory card that the device might be equipped with. Information that could typically be useful to spies is presented in Table 5.1.

A significant difference, when compared to computers, is that smartphones usually allow less control over what information is stored and where. This way, traces of applications and data remain in the memory and can be stored in areas that the user may not have access in order to delete them. Moreover, in many occasions, simply deleting a file [e.g., a Short Message Service (SMS)] does not lead to the immediate deletion of the element, but instead it merely flags the corresponding memory space as free and available to store new data. So, if that memory position is not overwritten by newer files, then the "deleted" files remain in the memory even though the user assumes they are deleted.

5.3 Cellphone Theft and Data Exportation

5.3.1 In General

The actual theft of a smartphone possibly constitutes the easiest way for violating the user's privacy. By stealing such a cellphone, a spy can extract information from the memory and the SIM card. Regarding the SIM card, it is possible to gain access by issuing commands to the card's microprocessor by using a smartcard reader (Fig. 5.1) or from the cellphone itself as long as it allows issuing such commands. In any case, specialized software makes the process easier, providing graphical interface and automation for a variety of tasks (Fig. 5.2).

Fig. 5.1 A simple SIM card reader

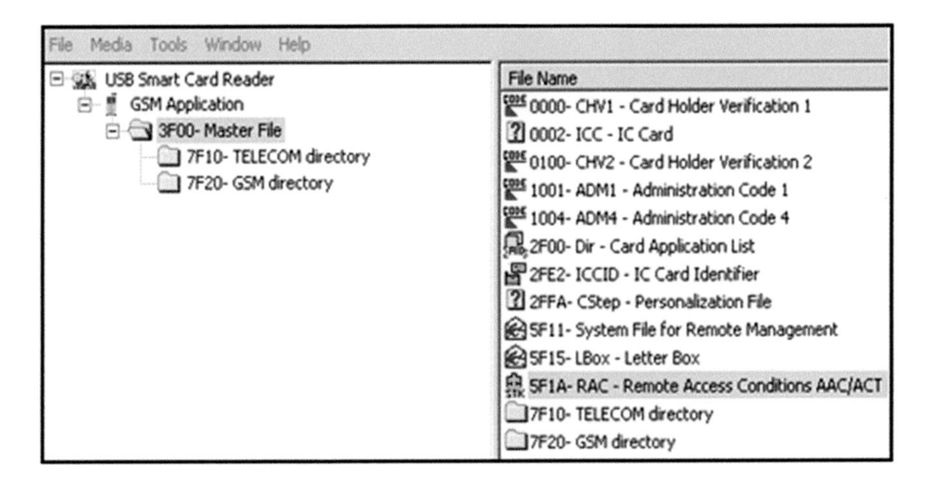

Fig. 5.2 Snapshot of the directory structure of a SIM card as shown using specialized software

5.3.2 *Files on the SIM Card*

The GSM 11.11 standard and its descendants require using a directory structure and specific files on the SIM card. Without going into more details, it is important to stress out that inside those directories there are memory areas and files where various information elements are held. Figure 5.3 shows that structure where MF (master file) is the root directory, DF (dedicated file) is the subdirectory, and EF (elementary file) is the file that contains the actual data. These files have different authorization levels for accessing, modifying, writing, or deleting them. Some of them can be read even without entering the PIN code; some others require entering the PIN code, while for the more important ones, only the provider has access by entering the appropriate password.

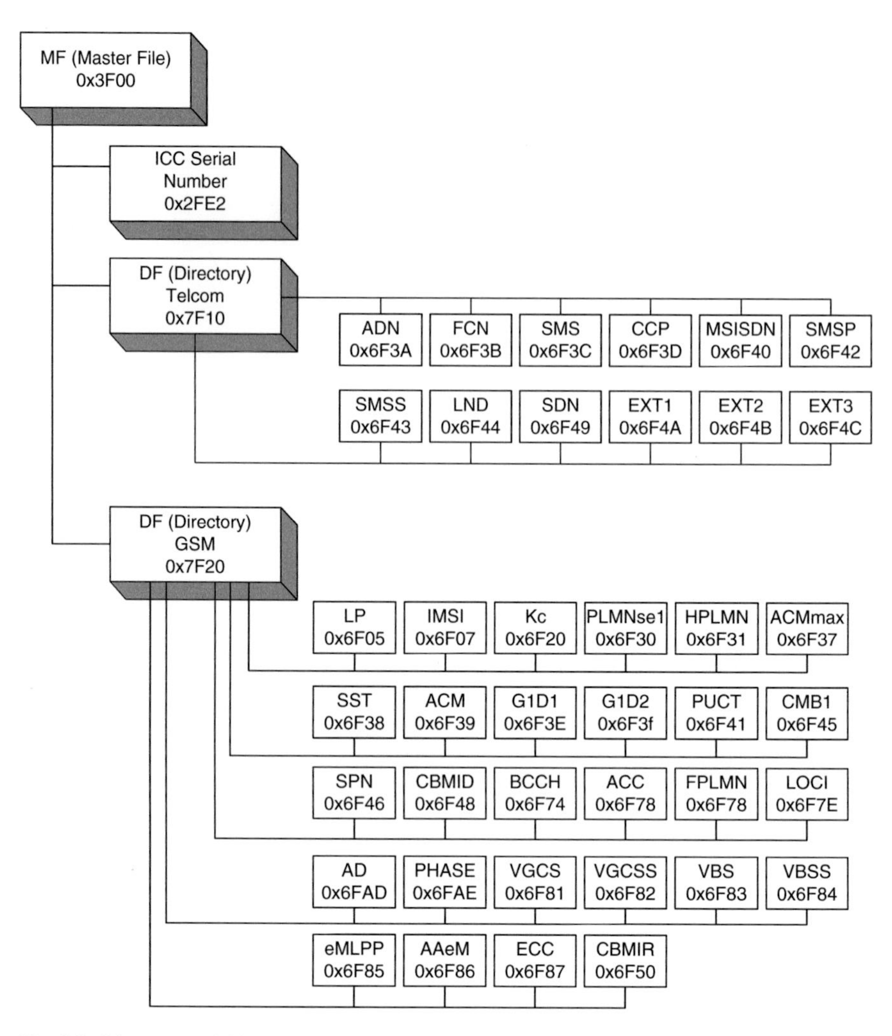

Fig. 5.3 Directory and file structure of a SIM card

For example, the SIM card contains files that include the card's serial number, a list of providers and their names, the preferred network, the preferred languages, the contacts' directory, incoming and outgoing SMSs, the settings for SMSs, and a list with recent outgoing calls. Also, there are files regarding the subscriber's identity on the network (IMSI-TMSI), the subscriber's location (LAI), the broadcast control channels (BCCH), the current encryption key (Kc), and dozens of other files (around 100 in total). It is particularly interesting that besides those standardized files, each provider often uses proprietary files. A brute-force reading process of all the memory addresses can reveal even more information beyond the information coming from the standardized files. A substantial part of the information included in these files (mostly the data regarding the network usage) is usually not directly accessible to the phone's owner. That way, a spy can extract important information that the user was not able to delete.

In the case of cellphone theft, the stored SMSs will also be available to the criminal. Moreover, the messages are stored on the cellphone's memory, but there might be some SMSs stored in the SIM card. Modern SIM cards have 35 slots for storing SMSs, but of course messages are also stored on the cellphone's memory when the available space on the SIM card is full or if the cellphone's manufacturer has chosen the cellphone's memory as the default memory.

The interesting point here is that deleted messages from the SIM card can be retrieved. For messages stored on the SIM card, the first byte of each slot for storing messages is indicating the state of the message and can have the following values (8 bits):

00000000 Unused
00000001 Read incoming SMS
00000011 Unread incoming SMS
00000101 Outgoing and sent SMS
00000111 Outgoing and unsent SMS

When the user deletes a message then it's just the value of that first byte that changes, while all other bytes where the message was stored remain intact. For an already read message, the corresponding byte has the value 00000001. If the message gets deleted, then the byte will get the value 00000000, but the message itself will continue to be stored at the same memory slot until it is overwritten from another message (since the memory slot was flagged as free). By directly reading that memory area, the attacker could read deleted messages. Luckily, some cellphones during the deletion of a message take one further step and fill in the slot that the message was stored with binary ones, rendering its later retrieval impossible.

5.3.3 Device Data

Despite the impressive capabilities of the SIM card, some data is important to be stored on the device's memory anyway. For that purpose, two kinds of memory are used, the NAND flash and the NOR flash. Data that are stored in the internal flash memory depend on the model/manufacturer but usually include the IMEI number, time settings, ringtone and volume settings, SMSs, calendar-alarm clock data, missed and received calls, extended directory with multiple fields, executable files and applications or games, and of course multimedia files such as pictures, videos, and recordings. Adding the ability to access the Internet, information that is stored on the cellphone further extends to include web addresses that the user visited and possibly their contents, favorite websites, names of Wi-Fi access spots, etc.

Apart from the current data, older deleted data can be found "deep in the memory" of the cellphone. Text messages, pictures, MMSs, simple or calendar notes, contacts, etc. are possible to be retrieved (partially or fully) after being deleted, under certain circumstances. Even copies of data from previous SIM cards may have remained in the memory (e.g., the IMSI number from the previous SIM card that was used on the same cellphone). A relevant forensics technique is that of file carving, that is, the process of reassembling computer files from fragments.

5.3.4 Data Extraction

The attacker wants to extract as much data as possible. For that reason memory dump techniques may be used in a byte-to-byte manner for NOR memory types or in a page-to-page manner for NAND memory types (different memory types have different access characteristics on the hardware level). This can be accomplished with specialized software and hardware tools that "clone" the contents of the memory. This way, a complete image of the data can be extracted on the physical layer. Technically speaking, the whole process is quite complicated since the data is unstructured and need to be "translated" to a specific file system.

These tools are available for various cellphone brands with the fundamental task of loading new firmware, updating, servicing, and debugging, and they are distributed through both official manufacturer channels and, unofficially, from third-party companies and individuals. A typical example of such a product is shown in Fig. 5.4. The unofficial distribution of such tools is usually related with illegal actions such as unlocking the phone (removing constraints the provider has set) or counterfeiting of serial numbers. In order to connect these tools to the device, no single common physical interface exists (each manufacturer uses different ones even between its own models), while they often require access to special internal interfaces (such as the JTAG interface). The decisive contribution of these tools to the data extraction process lies in the fact that they can operate even when a device is turned off, locked, jammed, broken, and so forth. Besides using hardware, there are various software collections (data suites) that allow high-level access (i.e., through the device's OS) to the memory.

If no other method can be used (e.g., the cellphone is partially or completely damaged), it is possible to detach the integrated circuits of the memory (Figs. 5.5 and 5.6) using specialized SMD (surface mount device) attaching/detaching workstations

Fig. 5.4 Cellphone read/write device. Various types of cables can be connected to the RJ45 port depending on the exact phone to be serviced

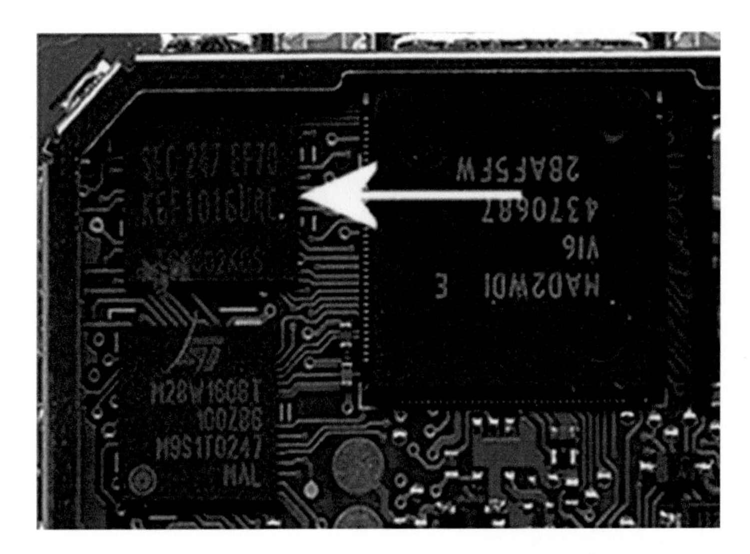

Fig. 5.5 The *arrow* shows the device's integrated circuit of the memory

Fig. 5.6 The contacts of the same integrated circuit of memory (μBGA type), after detaching it

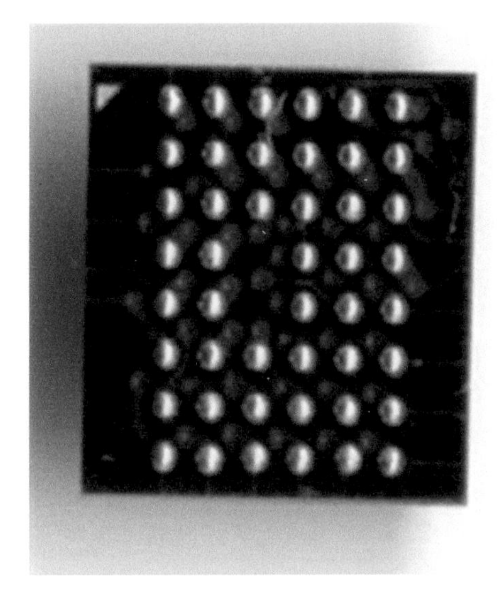

(Fig. 5.7). Then, the attacker can externally read the contents using the appropriate interconnecting equipment (Fig. 5.8). This process poses a significant risk since it can lead to the destruction of the integrated circuit due to the high temperature needed for detaching it. Moreover, the cellphone has to be dismantled in order to extract the integrated circuit of the memory. Due to the difficulty of the process, this method is the last resort for the attacker.

Fig. 5.7 Professional workstation for attaching/detaching electronic components

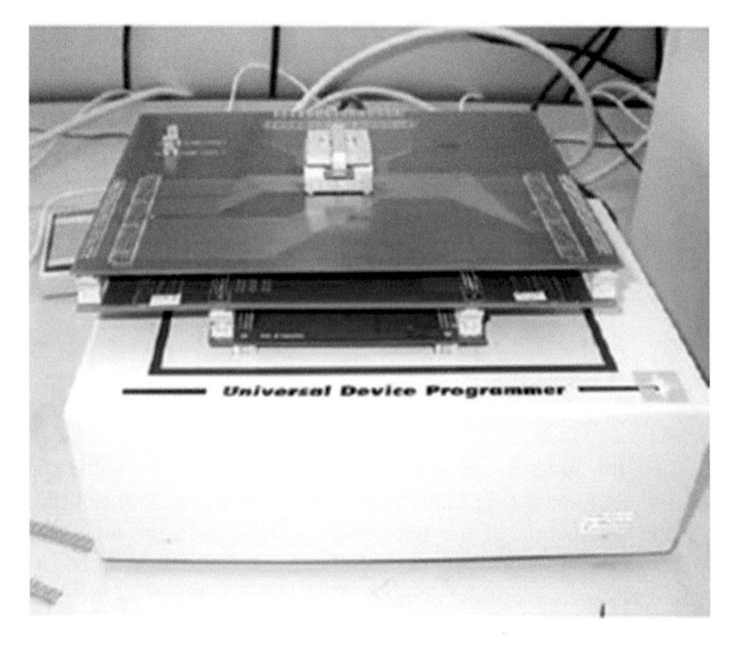

Fig. 5.8 General purpose memory integrated circuit reader

5.3.5 External Memory Cards and Computers

Due to the plethora of capabilities of modern smartphones, eventually the internal memory gets depleted. This is the reason why external memory cards have become very popular. They offer the capacity to store large volumes of data including videos, photos, songs, and other types of data. At the same time, they can be read from computers and other devices such as digital cameras and camcorders, being a very easy-to-use media. In order to connect memory cards, there is a variety of different interfaces and standards, but microSD is the dominant one. Indeed it is very easy to read the memory card's contents since there are dozens of products for that, for both novice and advanced users. Once again the contents of this memory can be extracted providing valuable information to spies. To conclude our analysis, it is worth mentioning that these memory cards can be connected to a computer, or correspondingly a computer may be connected to the cellphone (e.g., for creating backups). This poses excellent opportunity for spies to infect the victim's computer with malware. Imagine a cellphone left unattended for a few minutes. The attacker can load malware on the memory card and return it to the owner. That can easily take place at a gym or a restaurant. The next time the cellphone or the memory card is connected to the computer, the spy gets access.

5.4 Interception Software

As already mentioned, the theft of a cellphone can be only temporary, for as long as it's needed to be tapped with software or other electronic circuits so it can be transformed into an excellent informer that the victim will be constantly carrying around with him. Indeed, there are dozens of interception programs available that can be installed on a cellphone. Upon installation, these programs transmit to the attacker's cellphone (or to a server in the Internet) all the activity taking place on the tapped cellphone. Settings in a typical interception program are shown in Figs. 5.9 and 5.10. It is possible for the attacker to extract the device's contact list, intercept information (caller ID and time stamp) about incoming and outgoing calls, and have access to incoming and outgoing SMSs. More advanced versions allow eavesdropping on the ambient sounds and discussions. When the cellphone is in the office, for example, the attacker can silently activate its microphone and intercept conversations taking place close by. Of course, telephone conversations can also be intercepted in real time. When receiving or making a call, the program informs the attacker, and then he can piggyback the conversation by making a simple call. This method is based on three-way calling so there is a discrete tone heard at regular intervals to inform the participants about the three-way calling (according to what have been described in the previous chapter regarding landline interceptions). Unfortunately though, very few realize exactly what is going on and what the purpose of that tone is. To overcome this possible telltale, other spying software versions are recording the conversations in the memory of the phone and then silently uploading them in a server in the Internet, where the attacker has access and can download the recordings.

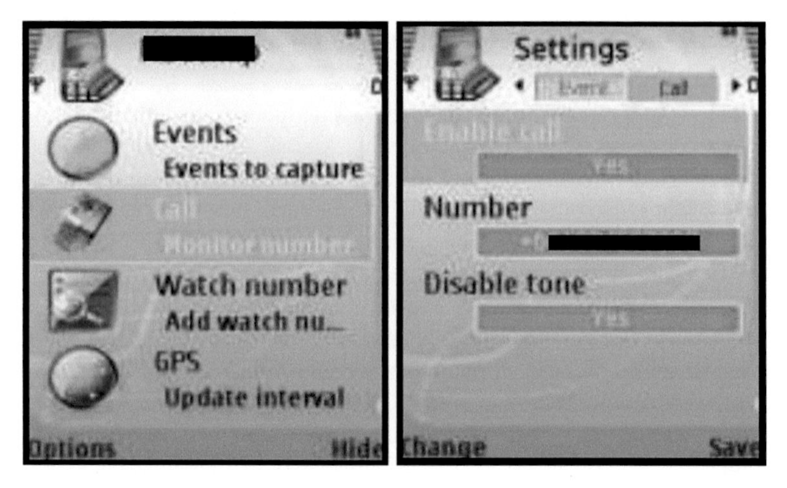

Fig. 5.9 Settings of a cellphone interception software

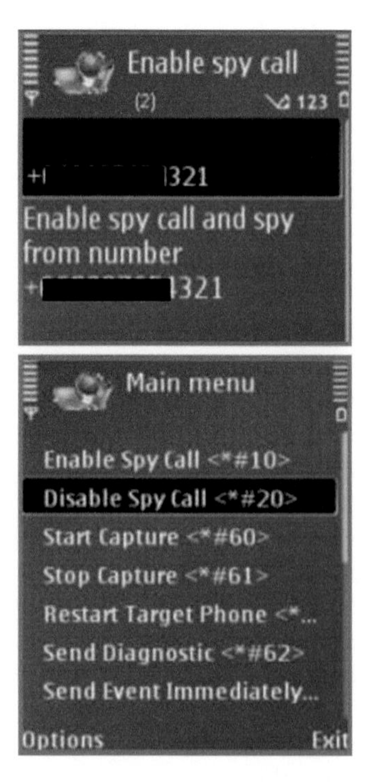

Fig. 5.10 Settings of a cellphone interception software

Such software can be installed within a few minutes as long as the attacker has gained physical access to the device. The attacker can also use social engineering techniques to persuade the victim to install such a "Trojan horse" unintentionally by advertising and promoting the software as a simple game or something similar.

Quite ironically, some antivirus and anti-malware programs are nothing more than such "Trojan horses." The user believes he has installed an antivirus while he has installed a malware.

Even without physical access to the device, the plethora of interfaces that modern smartphones are equipped with (i.e., Bluetooth, USB, Wi-Fi, GPRS, EDGE, HSDPA/HSPA+, LTE, and soon 5G) offers more attack entry points for both the device and the provider's network.

5.5 Network Security

Beyond devices, there are vulnerabilities regarding the cellular network itself. We continue with our analysis on that. As it is well known, GSM is a digital system for mobile wireless communications for voice and data which provides international access, high capacity, and transmission quality. Due to the multiple advantages of GSM, it soon defined the roadmap for mobile communications globally. One of the basic features of the standard was the encryption of voice which was a breakthrough compared to the analog systems used in the early 1990s. However, it was soon discovered that the protection offered by the system was not sufficient, and it was only a matter of time before the corresponding mechanisms and encryption algorithms were bypassed.

Since cellular networks are typical wireless communications, the radio waves propagate through the air freely, and they cannot be easily confined. Moreover, by confining their propagation we would automatically be affecting the range of the signal. So, a potential intruder could intercept and process these signals without having to be in the exact same location as the victim. To be precise, an attacker does not even have to come close since by using directional antennas, the effective attack range can be hundreds of meters or even kilometers.

By definition, the link of the cellphone with the network is of course wireless. Besides that link, there usually exist other wireless links that convey voice along with signaling transmissions originating from base stations located in remote and inaccessible areas that are being forwarded to the rest of the network via microwave links. Problems with such links were described in the previous chapter.

The GSM network does not only consist of wireless links but also of wired landline ones. Links between the switching centers that are located in metropolitan areas have no reason to be wireless since wired circuits are easily accessible in such areas. This constitutes another flaw in the initial designs of cellular telecommunications since they provided native security only to the cellphone access level to the network and not end to end during the rest of the transmission. Encryption usually stops at the link of the cellphone to the antenna at the nearest base station, and from that point onwards, the transmission of the data to the rest of the network is no longer encrypted. This usually happens for cost reduction reasons since encryption takes up bandwidth resulting in the reduction of the available channels used by the subscribers to place phone calls.

In the same manner, communication protocols between the switching center and the base stations (BSSAP) but also the communication protocols used in the core international signaling network (MAP-SS7) keep transmitting the data unencrypted. A skilled "insider" can intercept data that are being transmitted through the backbone network and then analyze the data to gain access to the context of phone calls and messages but also to the SIM card and encryption parameters.

5.6 Encryption and Cryptanalysis

For the GSM standard the "security through obscurity" method was selected to provide security despite the fact that the procedural guidelines for the implementation of some algorithms were publicly published. The vendors created proprietary "closed" solutions, avoiding to exactly describe how their algorithms operated. This choice leads to technical implementations that lacked the independent audits placed from the academic and the rest of the technological community. The results of that strategy were soon to be revealed since besides the theoretical analyses, practical attacks were also successful on various security levels within the GSM security framework.

Table 5.2 below presents the chronological order of the research outcomes and the successful attacks on the security algorithms. It is worth mentioning that the implementation of the GSM standard itself begun in the early 1990s, and its use peaked in the middle of the decade when the attacks described actually shattered the protection mechanisms in a very short time period. Furthermore, many of the attacks are based on passive analysis without any intervention on the provider's network itself. Such methods make the attacks "invisible," and it is impossible to be detected by either the provider or the user.

Table 5.2 Chronological order of attacks against the GSM security algorithms

April 1998 The Smartcard Developer Association (SDA) and scientists from UC Berkeley cracked the SIM card's COMP 128 algorithm and exported the Ki within a few hours of processing. They discovered that the Kc uses only 54 bits instead of 64 bits that was the theoretical key length
August 1999 The weak A5/2 algorithm was cracked within a few seconds using a simple computer
December 1999 Alex Biryukov, Adi Shamir, and David Wagner publish a paper where they succeed to crack the strong A5/1 algorithm. With a 2 min intercepted call, the attack time is just 1 s
May 2002 IBM's research department discovers two new methods (side-channel attacks) for intercepting the encryption keys from the COMP 128 algorithm
2003 Barkan and his colleagues conducted an active attack by transmitting data that could easily trick a GSM phone into using the weak A5/2 algorithm instead of the A5/1

(continued)

Table 5.2 (continued)

2006 Barkan, Biham, and Keller succeeded in the cryptanalysis of the whole A5/X algorithm family. The attack against the A5/2 algorithm needs only a few thousands of a second of an encrypted voice call to reveal the encryption key in less than one second on a simple PC A more complex attack against the A5/1 algorithm was published, involving an active attack and transmission of data. Also attacks against the network protocols that use the A5/1 and A5/3 algorithms or GPRS that takes advantage of flaws and mistakes in the protocols and are effective when the cellphone supports weaker algorithms like the A5/2
2007 The Universities of Bochum and Kiel started a research on producing massively parallel FPGAs under the name "COPACABANA" used for brute-force attacks without the need of extensive precomputed lookup tables
2008 The group "The Hackers Choice" started a project for a practical attack on the A5/1 algorithm by constructing a 3 TB precomputed lookup table
2009 A similar effort by Karsten Nohl and Sascha Krißler was based on the creation of precomputed lookup tables for the A5 using GPU cards. By December that year the first parts of those lookup tables that were calculated by a peer-to-peer network were made available through BitTorrent
2010 Improvements were made for porting the code in order to run on faster GPU cards
2011 Using precomputed lookup tables, a house computer equipped with a good GPU and some TBs of fast memory space could then crack the A5/1 encryption with a high success rate. Obviously, apart from the computer itself, equipment for receiving the signals is also required
2012 GSM security research has more or less stopped, with GSM security considered completely broken

5.7 Internal Fraud

Judging from the previous section, the encryption level of GSM cannot be considered as adequate under any circumstances. Let's assume for a moment that this wasn't the case, and truly strong encryption algorithms were implemented. Even in that case, the user would be facing a different kind of threat: attacks originating from within the network from malicious users and administrators.

We won't be analyzing frauds regarding overcharging since those don't fit the industrial espionage context. System administrators though can intervene and intercept data, voice call information, SMSs, etc. Abuse of the lawful interception mechanisms comes into the picture at this point. As it is known, all providers are obliged by law to have implemented such lawful interception mechanisms and provide them to law enforcement agencies when asked to do so. An employee with sufficient authorization or someone who can gain unauthorized access to this system, with or without insider help, can intercept at will both the voice calls and the SMSs of the victim he/she chooses.

Fortunately there are appropriate specialized software and monitoring and auditing mechanisms that can detect all the suspicious actions made from the provider's system administrators that we described above. In any case (and this applies to all information systems) using the appropriate security policies and clear separation of responsibilities can significantly reduce the risk and provide substantial security measures as well as de-escalation procedures for incidents once they occur.

5.8 Phone Call Interception

5.8.1 Introduction

Besides being a mobile computer, smartphones still serve the basic need of making phone calls. It is a fact that voice communications are probably the most immediate means of communication. Therefore, phone calls themselves are always interesting for an attacker. During the first analog mobile systems' era, intercepting phone calls was an easy process, and the necessary equipment was available to anyone that was interested (merely a radio-frequency scanner). The transition to digital systems like GSM changed the mobile communications' landscape by providing native encryption. At the same time, the technical complexity of digital systems required equipment that was both expensive and hard to find for active attacks to be launched so those systems remained secure for several years.

As it usually happens in every technology introduced in the telecommunication industry, the scientific community started theoretical approaches and research regarding the security of the encryption algorithms that were being used. Over time, those theoretical attacks were put into practice revealing new problems, and despite the various updates, the GSM protocol is no longer considered to be a secure (at all) protocol for mobile communications.

As it was shown on Table 5.2, in the international scientific literature but also on the Internet, there is a plethora of papers and information regarding practical attacks and cryptanalysis against the protocols themselves and their initial design. Besides those, there are various problems spotted on the implementation of those protocols from the various cellphone vendors or on the operation of the various subsystems and settings of mobile networks.

In this section we will be examining a specific and very successful type of attack which is based on using a fake, rogue base station for mobile telephony. Base stations are part of a mobile network that provides, by using the appropriate transceivers and antennas, the necessary channels connecting the cellphones to the provider's network. Even though the method to be shown pertains to GSM networks only, even the 3G/UMTS users are not immune to this attack as it will be shown later on.

5.8.2 Theory

One of the biggest security flaws and a core weakness during the GSM security design is the lack of provision—obligation of the network side to authenticate itself to the end user. As it is well known, in order to gain access to a provider's mobile network, the user needs to have the appropriate SIM card inserted in his device. The SIM card ensures the authentication of the user to the network by comparing the keys and identifiers that are contained in the SIM itself with the corresponding data that the provider keeps in its databases (the HLR—home location register). This way, the user can be authenticated before gaining accesses to the provider's services.

This basic security policy of GSM authentication though does not apply from the provider's side. That means that the active equipment of the base stations do not have a similar mechanism for authenticating their identity to the cellphone. Respectively, cellphones are unable to verify that the system they are connected to is legitimate and part of the actual provider's network that the user intends to connect. In that manner, the attacker only needs to activate a fake, rogue base station within a specific area and pretend that it is part of the legitimate provider's network that the victim is using. An aspect that makes the attacker's effort easier is actually one of the basic characteristics of the GSM protocol: every cellphone monitors the transmission of special data from the nearby base stations so it can connect to the one providing the best signal at any given time (usually the nearest one). This method is used so the power required for the transmission is kept to a minimum since by connecting to a nearby station, the cellphone would be transmitting with lower power thus increasing talk time. As such, if an attacker activates a high-power transmitter in a given area, overmasking the signals of the legitimate base stations due to the increased signal strength, then cellphones in that specific area will prefer to connect to the rogue base station. Indeed, now the rogue base station presents better characteristics than the legitimate one, so the mobile phones in the area will prefer it.

The next step of the attack focuses on disabling the encryption. In the GSM protocol, voice encryption is implemented using the A5 algorithm. There exist different versions of the algorithm providing different levels of security (A5/2, A5/1, and A5/3—weaker to stronger) as well as completely non-encrypted versions (A5/0). Under normal circumstances, the network keeps the unique cryptographic key (Ki) in the database of the authentication center (AuC) which is also stored in the user's SIM card and is never transmitted in the network. With the help of the A3 algorithm, those keys are compared and this process authenticates the cellphone on the network.

Following from the Ki along with other data and using the A8 algorithm, the session key (Kc) is generated which is used to encrypt the phone call with the A5 algorithm. In the case of the rogue base station, the basic information of the Ki key is unknown to the attacker so the attack shouldn't be able to continue any further. But for one more time, the design of the system gives priority to usability instead of security. So, the respective protocols allow the negotiation and the agreement on which encryption algorithm is going to be used (or whether encryption will be used at all).

With the appropriate signaling, the rogue base station informs the cellphone that it cannot provide encryption (it uses A5/0) so the cellphone is instructed to communicate with it without encryption. That way, the digital communication that will follow can be easily demodulated and recorded immediately.

Another characteristic that makes these attacks easier is the fact that those attacks are usually targeted, in the sense that a specific victim is being monitored. So, a few milliwatts of transmission power are enough for the victim's cellphone to connect to the rogue base station placed 20–30 m away rather than to the legitimate provider's base station located, say, 500 m away since the first one provides stronger signal level.

Up to this point we described the rogue base station's methodology for "capturing" the cellphones located within its range. But for the phone call to go through, a connection to a legitimate network is further needed. The interested reader will recall that implementing a man-in-the-middle attack plan is sufficient to intercept the phone call. Indeed, the attacker connects the rogue system with the rest of the network through a simple cellphone or a landline which will relay the communication to the legitimate network and toward the initial call recipient. Needless to say that apart from relaying the communication, the attacker will be intercepting and recording it as well. Figure 5.11 represents the exact topology of the attack. As it is shown, the two cellphones located near the rogue base station are under control of the attacker and are only communicating with the legitimate network via the relay cellphone. On the other hand, the four cellphones in the middle and the two in the left part of the figure remain unaffected since they are outside the reach of the fake base station.

It must be noted here that third generation (3G/UMTS) and further networks provide mutual authentication between the cellphone and the network infrastructure so at first glance this attack does not apply. However, cellphones by default (the user can luckily change this behavior via telephone's settings) will connect to any available band, since once again usability is given precedence over security. Therefore, the attacker can launch a jamming attack (jammers were discussed in a previous chapter) in the appropriate frequency band used by 3G networks. This way the attacker would be disabling 3G coverage forcing the cellphones in the nearby area to fall back to GSM mode. By looking for networks on the GSM frequency band to continue their operation, the target phones will inevitably get connected to the attacker's fake base station. As such, with this method, the attack as described above can be successful without requiring any further actions, effectively mitigating the security measures of 3G.

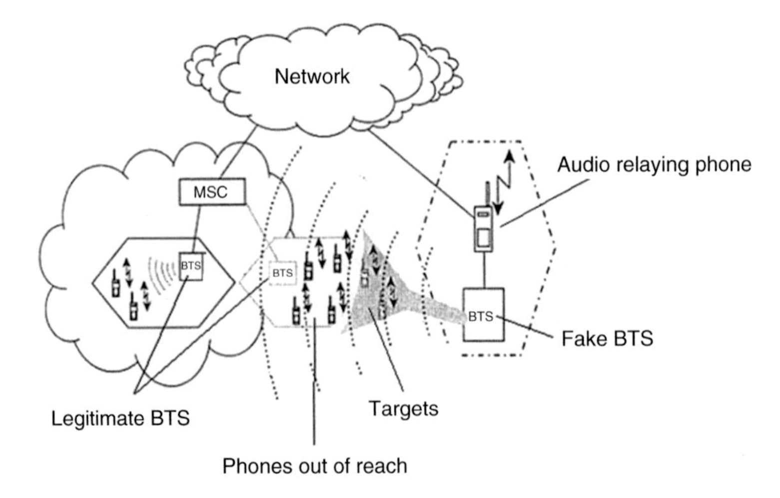

Fig. 5.11 Schematic of the attack

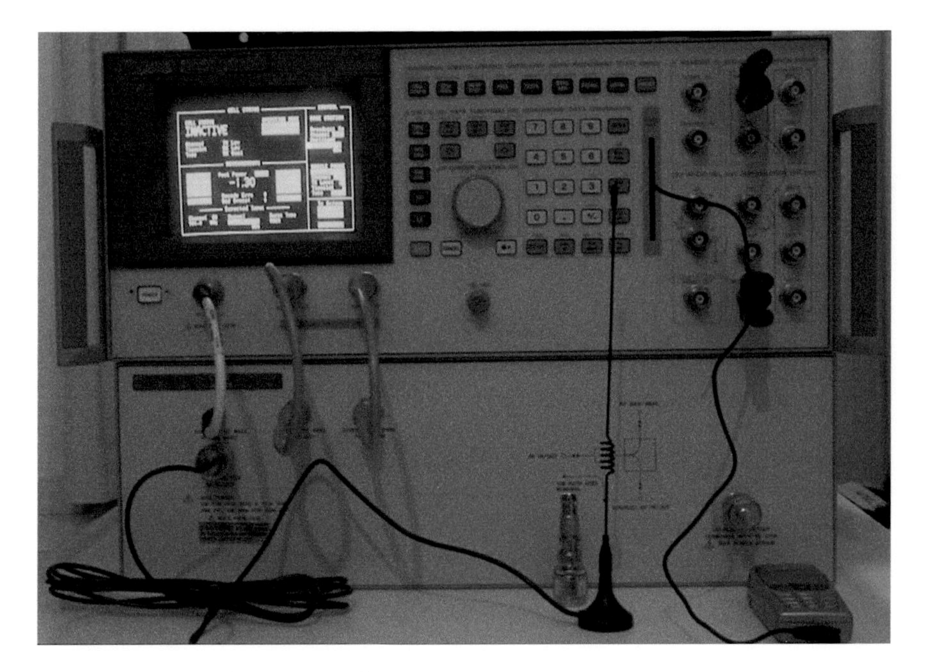

Fig. 5.12 Cellphone diagnostics device

5.8.3 Implementation

The ideal setup for deploying such attacks would of course consist of an industrial base station like the rest of the base stations in a typical mobile network. Such a setup though would be too expensive and too difficult to transport. An easier and cheaper alternative for such attacks would be to use advanced diagnostics and repair devices for mobile phones (GSM testers). Such a device is shown in Fig. 5.12.

Equipment like that provide all the necessary signaling to the cellphone exactly like a base station would do and at the same time receive and demodulate the digital signal and the speech that is transmitted. During the connection with the cellphone, the equipment conducts various measurements of the link characteristics so it is possible to detect any malfunction that the cellphone might be experiencing (Fig. 5.13). Even more interesting, they can place phone calls to the cellphones while being able to choose the caller ID that will be shown up. Besides voice calls, those devices can also send and receive SMS messages from the cellphone that is under their control. More advanced techniques allow sending SMSs in binary form (binary SMS) which can be used to send source code (e.g., viruses) and firmware updates without the user being unaware of the process. Such SMSs are not possible to be send over the legitimate network since each provider can detect them and block the request if they are originating from an untrusted source. In any case, facilitating the attacker, the configuration and the operation of such devices can be automated by connecting the testers to a computer through the appropriate interfaces.

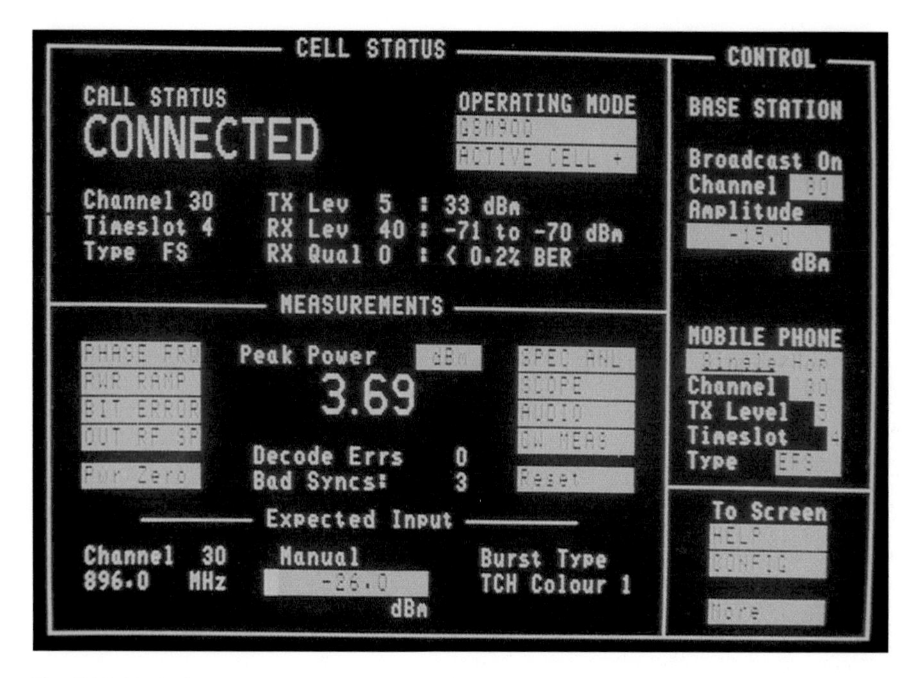

Fig. 5.13 Typical measurement taken by the cellphone diagnostics device

Finally, as shown in Fig. 5.14, rogue base stations can extract the telephone number that the cellphone under their control intends to call, the subscriber's/victim's unique ID that is stored in the SIM card (IMSI number—International Mobile Subscriber Identity), as well as the unique number that identifies each cellphone (IMEI number—International Mobile Equipment Identity). The IMSI number uniquely identifies a subscriber in a global mobile telephony network and along with the Ki key (which can also be extracted by the rogue base station by repeatedly querying the cellphone to perform cryptanalysis—brute-force attack method) can be used to clone a SIM card. In the same figure, it is worth noticing the encryption option field (ciphering) being empty, exactly like the attackers' wants, since he does not pose the encryption keys.

As it was previously mentioned, those devices are intended to be used for diagnosing and repairing cellphones. They are usually connected to the cellphone to be examined via a specific cable. That would be enough to limit their malicious usage, but unfortunately that limitation can be easily overpassed. Instead of a cable, the attacker can connect a signal amplifier setup and drive the transmitting signal to an antenna. Such amplifiers (Fig. 5.15) are very easy to use since besides the power supply they only need a connection cable to the diagnostics' device and an external antenna.

There are different types of amplifiers which, depending on their output power, can cover progressively bigger areas. In practice this would not be desirable since in a wider coverage area, the attacker would "capture" a bigger number of cellphones. This would render tracking the target more difficult and would lead to resource exhaustion for processing the data. Indeed, these diagnostic devices—GSM

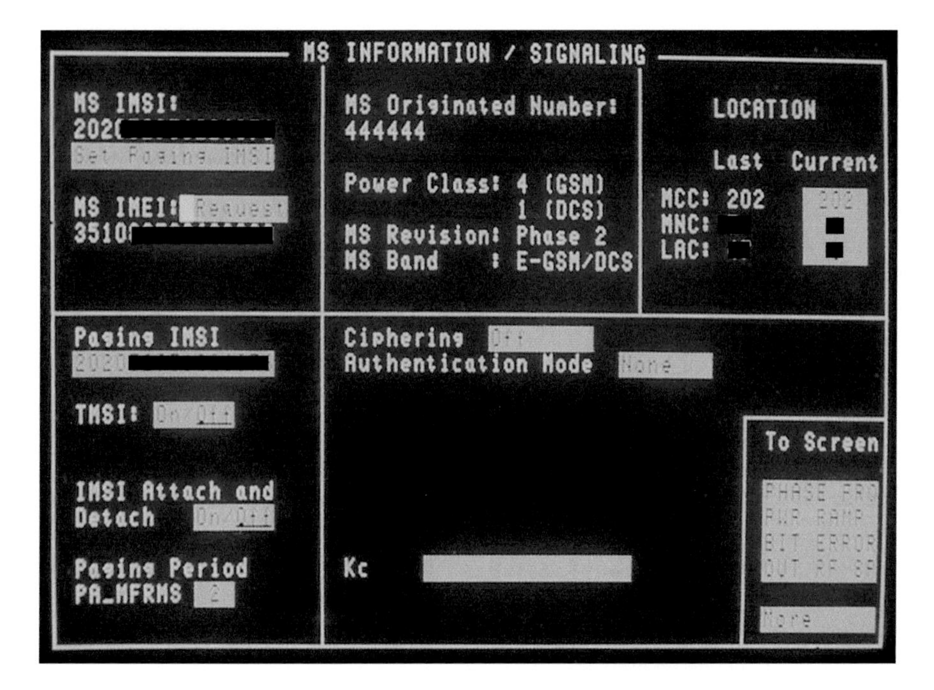

Fig. 5.14 The rogue base station has extracted the IMSI number, the IMEI number, and the phone number that the victim's cellphone is trying to call (444444)

Fig. 5.15 GSM signal amplifier

testers—are usually equipped with one or a maximum of two transceivers so they can simultaneously service one or two calls, respectively. So in case the attacker tries a large-scale attack, these limited resources along with the problems that will be caused to the network would quickly reveal his existence. Of course, at the end of the day, everything is relevant to the equipment's cost and to the amount of time that the attacker is willing to spend since there are more advanced devices that are equipped with multiple transceivers allowing for eight, 16, 32, or even more mobile phones to be tracked and intercepted.

5.8.4 Configuration

After the appropriate connection of the equipment, some configuration needs to take place. The necessary information that needs to be configured on the equipment is initially the mobile country code (MCC) and the mobile network code (MNC). It is worth mentioning that the ITU E.212 standard (Land Mobile Numbering Plan) defines all the MCC numbers to be used on the mobile telephone networks thus making that information publicly available (Greece, e.g., is appointed MCC id 202). Then the channel number (frequency) where the network synchronization and control signal will be transmitted (BCCH—broadcast control channel), as well as the channel number for transmitting voice (TCH—traffic channel) through which communications will take place, needs to be entered. Finally, it is important to deactivate the encryption according to what has been described earlier. Further to those basic settings, a plethora of other parameters (Fig. 5.16) are available to be configured for more advanced attacks.

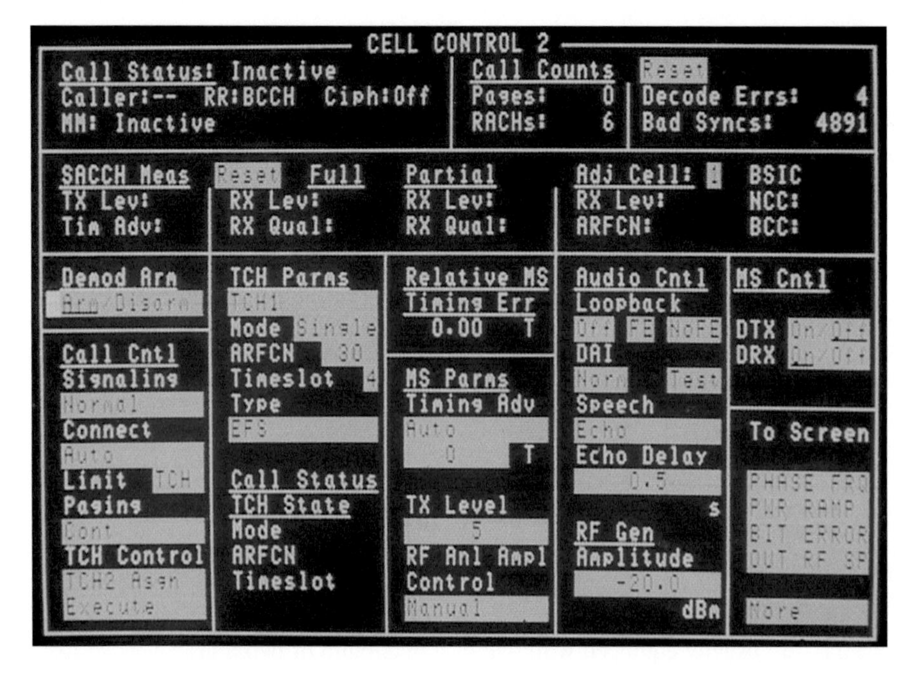

Fig. 5.16 Advanced settings

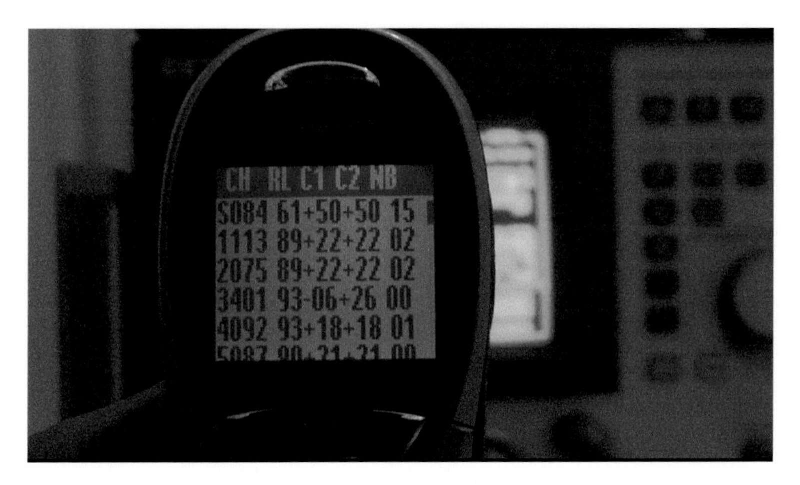

Fig. 5.17 Information about the available channels (ARFCN) and the received signal power as displayed using the Netmonitor software

Finding the appropriate parameters the legitimate network is using can be easily done by using specialized software such as the Netmonitor (Engineering Menu) software. Netmonitor is a specialized diagnostics software which allows its user to get detailed information about the operational parameters of the network, the cellphone, and the SIM card, while it even allows altering some of those parameters (the ones that pertain to the phone, not to the network). Figure 5.17 shows information provided by the Netmonitor software about the channels-frequencies (ARFCN—absolute radio-frequency channel number) and the corresponding signal power received from the transceivers of the nearby base stations. As already mentioned, the cellphone monitors those channels in real time, in order to choose among them the one that has the highest signal power. The attacker, therefore, configures the rogue base station to transmit on one of those channels with enough signal power to overlap them, effectively taking control of the nearby cellphones. Indeed, a few moments later, the target cellphones will be under the rogue base station's control which can even issue an appropriate command to prevent the connection with other channels-frequencies as it is shown in Fig. 5.18. As it can be seen there, the fake base station is not advertising any other nearby channels so it can keep the target mobile phone camped on to him for as long as it can provide the minimum signal strength necessary. Otherwise, the mobile phone could jump back to a legitimate base station at some point.

From this point on, the interception takes place, in accordance to what we have described, using the man-in-the-middle attack method. Following, we will not be expanding any further on this subject but we will be examining security issues regarding the basic service of SMS as well as the Bluetooth connections.

Fig. 5.18 The rogue base station is not recommending any other channels

5.9 Short Message Service

Short messages are the most popular GSM service after the basic telephone service. Unfortunately, besides the immediacy, ease of use, and effectiveness that this service offers to users, it also poses several threats.

As it was mentioned for voice communications, the encryption (if applied) stops at the point that the message enters the provider's network, and from that point onwards, the messages are forwarded in the core network without any encryption, following the mandates of various protocols depending on the vendor. So employees that have access to the corresponding provider's systems can, at will, have access to the message's context as well as information about the sender and the recipient.

On the other hand, an attacker doesn't necessarily need access to the backbone network. The malware we described earlier in this chapter, being installed on a cellphone, can retransmit all the messages being sent and received using that device to a third person. Of course, the messages being relayed from the victim's cellphone to the attacker's cellphone will be charged. So this activity along with the telephone number that was used to receive the intercepted message will be revealed with a closer look at the next bill. But it will be too late by then.... In any case the server-based malware, which uploads intercepted SMSs to a server in the Internet, can't be as easily traced since it does not leave a visible trail such as the phone number of the previous case.

At the same time, the rogue base station that was described in the previous section has the ability to intercept every message that the user tries to send along with the recipient's phone number. Since the cellphone is not connected to the legitimate network while being under control of the rogue base station, the intercepted message would never reach its destination. But since the attacker is aware of both the message itself and the recipient, he can send a copy of the message to the original recipient. The attacker can also change the originator's identity and choose to use the original sender's phone number in order for the interception not to be revealed. This is particularly easy through a number of Internet applications and bulk SMS providers. Combining it with social engineering techniques, the attacker can send a message while appearing to be a colleague, a technician from the victim's Internet service provider (ISP), etc. tricking a call recipient that only relies on the caller ID to identify the person who is calling.

Finally, there are techniques that can be used to retrieve already deleted messages from the cellphone's memory and from the SIM card as we have previously mentioned, although those processes depend on the different cellphone models, and they aren't always successful.

5.10 Bluetooth®

5.10.1 In General

Bluetooth®, which has been widely known from the headsets available in the market, is an industrial standard for wireless personal area networks (WPAN). It concerns a short-range wireless telecommunications technology that provides standardized communication between digital devices through a globally available and non-licensed radio-frequency band.

Besides some basic security and encryption features, attacks against cellphones (and not only cellphones) via Bluetooth® exploit technical details in the protocols and design flaws but mostly vulnerabilities and poor implementations from the various vendors. That is, while the standard might be secure enough, a reckless implementation opens the door for various vulnerabilities. At the same time, user behavior (mostly due to ignorance and lack of information security awareness) can lead to the cellphone being fully controlled by an attacker.

5.10.2 Data Interception via Bluetooth®

Any protocol used for communications and data exchange can in certain occasions lead to data interception. In our case, in certain older cellphones, due to poor implementations, it is possible to receive data through channels that were intended to be used only for sending data (e.g., sending contacts). Using these open channels it is not possible to browse the contents of the cellphone's memory, but it is possible to directly query specific file names. On these devices, the contact list, the calendar data, the stored pictures, and many other files have specific file names due to the standardization that took place during the infrared communications' bloom era (IrMC specifications). For example, the contact list is stored in the telecom/pb.vcf file, so a direct query to receive this file through the contacts' sending channel (which normally shouldn't be allowing requests for receiving files) is enough for the file to end up in the attacker's hands within a few seconds.

More straightforward interception techniques are based on the fixed PIN codes that Bluetooth® hands-free headsets have and can intervene in the connection, recording the call or even transmitting content different than the original. Indeed many handsets have default or even fixed PIN codes such as "0000," "1234," etc. As such they are effectively known to the attacker. It is also possible to intercept ambient noises, including conversations taking place in the area around the hands-free headset without it necessarily being in use, by activating its microphone.

5.10.3 *Targeting Users*

As it is known, devices can be in three different states when it comes to Bluetooth®: disabled, enabled but invisible, and enabled while being visible to everyone. So, a simple scanning is enough to reveal potential targets. At this point, it is worth mentioning that there is a significant (but reasonable) misunderstanding among users. It is generally believed that a user is sufficiently protected when the protocol is enabled but set as invisible. At this state the device may not be responding to scanning requests from other devices, but it does respond to direct queries on its unique address.

Indeed, every Bluetooth® device has its own unique address, the MAC address like network interfaces have on PCs. The MAC address consists of 48 bits that can be represented in a more understandable form for humans, e.g., 00:01:E3:21:4B:55. The first 24 bits state the vendor and are issued from the IEEE in accordance to the OUI catalog (Organizational Unique Identifier).

Respectively, the rest of the 24 bits are issued randomly from the vendor even though in certain occasions it is possible to associate them with specific models. In Fig. 5.19 some addresses like that are shown, with their last digits hidden.

So, if the MAC address of a specific device is known in advance, it is possible to connect to the device even when the protocol is set to invisible mode (just enabled). In any case, in order to pair two devices, both of them need to be in visible mode at least during pairing, so for that period the devices remain vulnerable. Fortunately some devices do not allow being in active mode for more than a specific time period (e.g., 5 min). This way they can effectively mitigate the risk of another device being paired with them involuntarily.

5.10.4 *Integrity (Cellphone Control: Billing)*

Unfortunately the problems are not limited to the passive data interception but can also have an active perspective. It is puzzling why there used to be older cellphones (some of them particularly popular at the market) that had hidden and unprotected channels, accessible via Bluetooth®, similar to the ports that operating systems have on PCs. Through these channels, it was possible to have full access to all the cellphone's functions without been prompted for any kind of password.

Fig. 5.19 Typical Bluetooth® device's addresses

```
00:1D:FD:72:
00:13:6C:6E:
00:12:D1:8E:
00:10:20:E8:
00:10:20:E8:
00:13:6C:01:
00:E0:0C:50:
00:13:6C:36:
00:23:D6:5A:
00:23:D6:5A:
00:1C:9A:25:
00:1E:45:80:
00:23:3A:B4:
00:1E:45:80:
00:10:12:E8:
00:1E:45:80:
00:14:A7:6A:
```

For that purpose AT commands were issued through serial connection over Bluetooth®. So, a malicious user could activate the call forwarding function, send messages, initiate calls, and of course intercept all the data stored on the cellphone.

Fortunately, cellphone vendors stopped making such unacceptable (or maybe deliberate?) mistakes. On the other hand, the embedded and general purpose operating systems found in modern smartphones brought back those threats in a different form (e.g., viruses and malware). Furthermore, besides cellphones, it is also possible to hack into a computer using Bluetooth® as long as there are flaws in the implementation of the embedded protocol using buffer overflow attack techniques.

5.10.5 Bluetooth® and Social Engineering

Apart from the potential denial of service attacks, sometimes the user himself is forced to turn off his cellphone. We are referring to one of the negative aspects of the popularity of the protocol which is harassing nearby users with the exchange of (free) messages via Bluetooth®. Initially, one had to set the device name to the text that he wanted to send (like in Fig. 5.20) and initiate the connection process with a nearby recipient. Alternatively he could create a contact with the same context as the name thus showing the message to the recipient's screen. Some models, even today, show the contact details (same as the message that the attacker wants to send) without any confirmation. There even exist applications that can automate the whole process of sending harassing messages via Bluetooth®.

Fig. 5.20 The device's name is set to "You're being watched" thus showing up during the pairing request

Besides annoying, this harassment can also be dangerous in the industrial espionage aspect if used in combination with social engineering techniques. Imagine the case where the malicious user sets the device name to PIN1234, CODE1234, Enter 1234, or something similar and initiates a pairing request. The victim's cellphone shows that message (as the ID of the attacker's device) and prompts for PIN confirmation. Then the unsuspected user that is under attack might enter what shows up on the screen (without realizing that he is actually typing a PIN that is already known to the attacker) unwillingly giving him full access to his cellphone.

Insufficiencies and flaws on the design of the interfaces is another significant issue. The standard provides a PIN with a maximum length of 16 bytes (thus including characters too) for the generation of a Bluetooth® key. The majority of cellphones, however, only accepts numbers and even of limited length. This insufficiency, regarding the design of the interface, significantly lowers the key strength. In addition to that, peripheral devices such as mice or hands-free headsets are usually found on the market having a fixed, factory-set PIN which cannot be changed since there is no easy way to input data on those devices. Indeed, even when changing the PIN is an available option, doing so requires using complicated microswitches or connecting the device to a computer or using a previously authorized Bluetooth® connection. Even worse, some cellphones allow files to be received without any confirmation from the user at all, while others only require a YES/NO confirmation for receiving data without enforcing using a PIN. Remaining on the interface design flaw subject, some ridiculous occasions have been observed where the cellphone gives up prompting for a PIN and allow the transaction, when the user cancels the request or when he enters the wrong PIN for 4–5 times in a row.

5.10.6 Tools (Hardware-Software)

A malicious user that desires to exploit the vulnerabilities that we described above has a variety of tools available at his disposal, both software and hardware ones. For the more determined one, there are Bluetooth® analyzers capable of monitoring the data packets transmitted during the devices' operation in real time and in a completely passive way over the air. In combination with cryptanalysis techniques, this method can reveal the context of the Bluetooth® communication. On the other hand, the cost for such analyzers is significantly high since they are intended for commercial products' quality control and mostly used during fabrication processes. In the past few years, though, there have been attempts for creating such tools using open-source code and cheaper general purpose hardware.

5.11 Protection Measures

Having described the various vulnerabilities and security flaws, we will be concluding this chapter with mentioning protection measures against voice and data interception in cellphones.

5.11.1 Protection Against Interceptions

According to the individual technical standards of GSM, which are known to cover thousands of pages of technical documents, there is one bit in a specific area of the SIM card (in the EF_{AD} table file) which if set to "1" it allows the cellphone to bring up the necessary user warning mechanism regarding the turning off of the encryption. That particular bit is called OFM (Operational Feature Monitor) and is described in Fig. 5.21. That bit was renamed into "Ciphering Indicator" in later versions of the standard in order to be more understandable. It is important to note that this bit is configured on the SIM card by each mobile network provider and can also be altered (usually deactivated) with the appropriate binary SMS as it was mentioned previously, without the user being aware of it.

The warning mechanism is in fact a small icon like the one highlighted with the red arrow in Figs. 5.22 and 5.23 (e.g., asterisks and an exclamation mark or an open padlock). Some vendors take it one step further by displaying a warning message for some seconds (Fig. 5.24). Unfortunately some other vendors choose not to implement this mechanism resulting in the user not being notified at all regarding the absence of encryption even if the corresponding bit in the SIM card is active.

Further to that mechanism, a user can enable/install the Netmonitor application on a cellphone and get informed about the provided encryption level (and whether encryption is switched off). This solution, even though not very practical, is unmistakably accurate since it depends on the network itself and shows the encryption level in any given time, regardless of the OFM bit value.

Identifier: '6FAD'	Structure: transparent		Mandatory
File size: 3+X bytes		Update activity: low	
Access Conditions: READ ALW UPDATE ADM INVALIDATE ADM REHABILITATE ADM			

Bytes	Description	M/O	Length
1	MS operation mode	M	1 byte
2 to 3	Additional information ←	M	2 bytes
4	length of MNC in the IMSI	O	1 byte
5 to 3+X	RFU	O	(X-1) bytes

Byte 3:

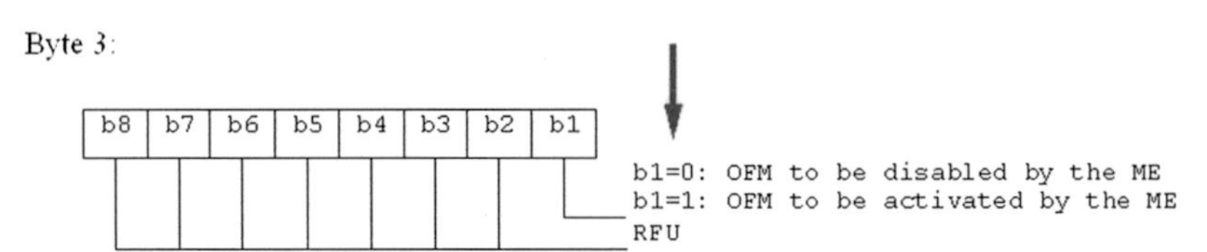

Fig. 5.21 Description of the location of the OFM bit (Ciphering Indicator) from the respective standard

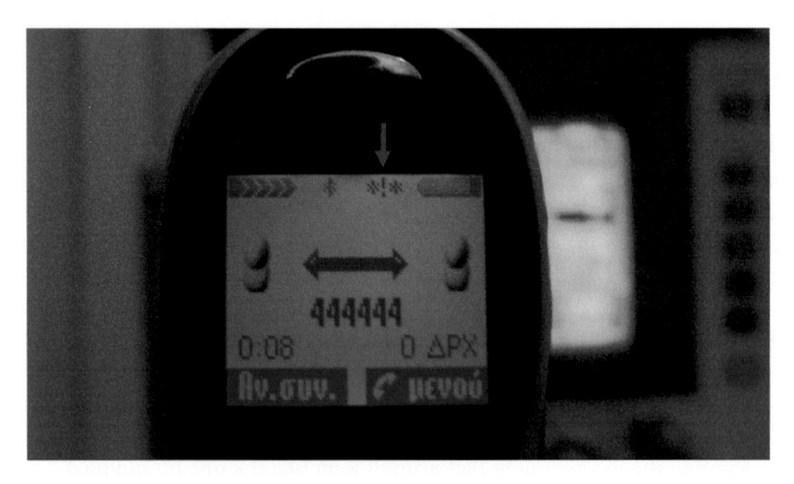

Fig. 5.22 The *asterisks* with the *exclamation mark* in the *middle* informs (those who are aware of their meaning) about the lack of encryption

The most basic parameter that can reveal a rogue base station attack is the isolation of the cellphone from the legitimate network. Indeed, during this attack the user would not be receiving incoming calls or text messages since the phone is out of the reach of the provider's network. The voice quality could also be deteriorated with delays and/or

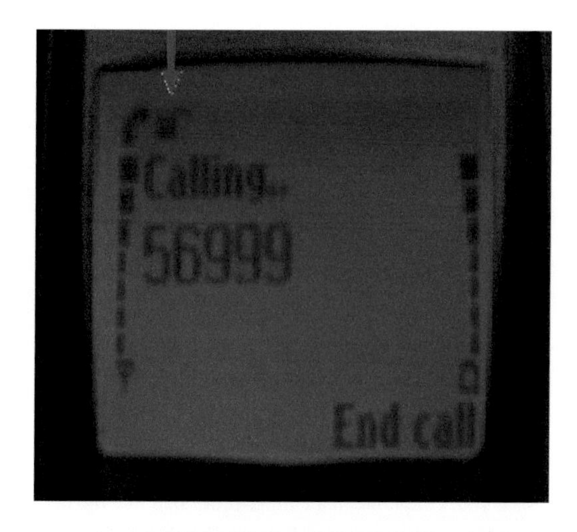

Fig. 5.23 The *open padlock* shows (a bit better than the last figure) the lack of encryption

Fig. 5.24 A significantly better implementation is the warning *triangle* with the *exclamation mark* which stays on during the entire length of the call in combination with the clear warning message that shows up for a few seconds after the call is initiated

echo. These clues along with a potential warning such as the one described above should immediately raise the user's concern about the security of the communication.

In order to eliminate this kind of interception risks, the network's authentication toward the device should be enforced. Fortunately, third-generation mobile network systems do solve this problem by implementing the network authentication feature. As it was mentioned though, by using jamming devices in the 3G frequency bands, cellphones in that area fall back to GSM operation so they become vulnerable once again. Besides that, as it usually happens in the technology sector, it is only a matter of time before weaknesses are revealed regarding modern networks and for new attacks to begin showing and affecting them.

5.11.2 Practical Advice for Cellphone Protection

Table 5.3 that follows includes practical advice for tackling most problems that we have mentioned so far. Following these guidelines a user can keep security on a satisfactory level making every attempt for espionage, industrial or not, much harder.

Table 5.3 Practical security guidelines for cellphones

Enable PIN code prompt on the SIM card
Keep the PIN code and the rest of the passwords secret
Enable the password protected screen saver lock
Do not lend your cellphone, not even for a few minutes
Do not let your cellphone unattended in public areas
If you are obliged to hand in your cellphone (e.g., during a visit to high security areas), first place your cellphone in a tamper-proof bag like the ones used for purchasing liquids at airports
Personalize your device so it cannot be changed with an identical one
Do not bring your cellphone to important meetings or remove the battery
Note down your device's IMEI number for future reference
Do not store sensitive data in your cellphone without using encryption
Keep your device's operating system and software updated
Get familiar with the icons on the screen and check them frequently
Enable the Netmonitor feature
Be careful in case the cellphone's screen turns on and off with no reason to
Get familiar with the calling tone, waiting tone, and the three-way calling tone of your network
Deactivate GPS when it is not needed
Do not visit unsafe websites from your cellphone
Do not accept unknown files through Bluetooth, e-mail, and MMS
Do not install unknown applications
Check the bill thoroughly for messages to unknown numbers and Internet usage
Install antivirus software if it supported on your device
Consider using crypto phones and applications for calling encryption, as well as for encrypting the data stored on the device
Consider using applications that provide remote device swipe in case of theft
Do not rely only on the caller ID to identify the caller
Keep Bluetooth disabled or at least in invisible mode
Do not accept Bluetooth pairing requests from unknown devices
Use long passwords for Bluetooth pairing
Do not initiate the pairing process in unsafe areas
Periodically check the Bluetooth-trusted devices list for the presence of unknown cellphones/devices
Pay attention to the characteristic sound from nearby loudspeakers when the cellphone is in use. If the sound is audible for more than few seconds when the device is not in use, there is the possibility that a silent call is being placed (if the humming sound is audible periodically for a few seconds every few hours or so, it is part of the normal operation of the cellphone—network update)
Periodically check the wireless access points, which are stored in the respective menu on the device, for the presence of unknown entries
Do not enter your phone number on websites (for games, ringtones, logos, contests, etc.)

5.12 Conclusions

The continuous growth of storage space in smaller dimensions allows the downloading, storing, and processing of more and more data on mobile devices and smartphones. In case of smartphone theft, the data stored in it can be used for industrial espionage. Respectively, using data and phone call interception techniques, cellphones are turned into a valuable ally for spies providing confidential data. As it was presented in this chapter, a malicious user with the appropriate equipment only needs to know a very few details of the network in order to intercept phone calls and text messages from nearby cellphones he targets. The attacker can also call the victims or send them text messages using any identity he may choose, while if he has enough time at his disposal, he can even clone/copy the victim's SIM card.

Like most of the security issues, the problem is not exclusively a technical one. Users themselves are not aware of the threats that might put them in danger and do not follow the best practices in order to protect themselves. Furthermore, they are not aware of the different functionalities and the (often unclear) indications of the devices they use on a daily basis (e.g., the indication for the absence of encryption that exists only on some cellphones). So, the problem is not purely technical as it was described so far, but it extends to the user awareness issue. Training and raising awareness is needed. At the same time, vendors and network operators ought to provide every other possible technology that is available for improving the security level of the stored and transmitted data. Such an attempt is described in the research paper "Androulidakis, I., & Kandus, G. PINEPULSE: A System to PINpoint and Educate Mobile Phone Users with Low Security. Proceedings of 7th International Conference in Global Security, Safety and Sustainability (ICGS3), Lecture Notes of the Institute for Computer Sciences, 99, 62–66, 2012," as follows.

Since users exhibit different levels of security feeling in regard to mobile phone communications, and since there are categories of users that face increased security risks due to their self-reassuring feeling that mobile phones are secure per se, a research proposed in the aforementioned paper describes a system that pinpoints and informs mobile phone users that have a low security level, thus helping them protect themselves. The system would consist of software application, installed in mobile phones, as well as of software and databases, installed in the mobile telephony operators' servers. Mobile telephony providers (by adopting this application), as well as manufacturers (by pre-installing it in their phones), could help mitigate the increased security threats effectively protecting the end users.

As Fig. 5.25 depicts, the system would consist of an application installed in mobile phones and software and databases installed in mobile operators' main servers. These applications communicate through the mobile telephony network in a ciphered way. The mobile phone-installed application (with minor differences in the array of services offered) would be able to function in all kinds of devices that have an advanced operating system (e.g., Windows Mobile, Symbian, Android, iOS). A lighter version could also be implemented for older and simple devices using J2ME (Java 2 Micro Edition).

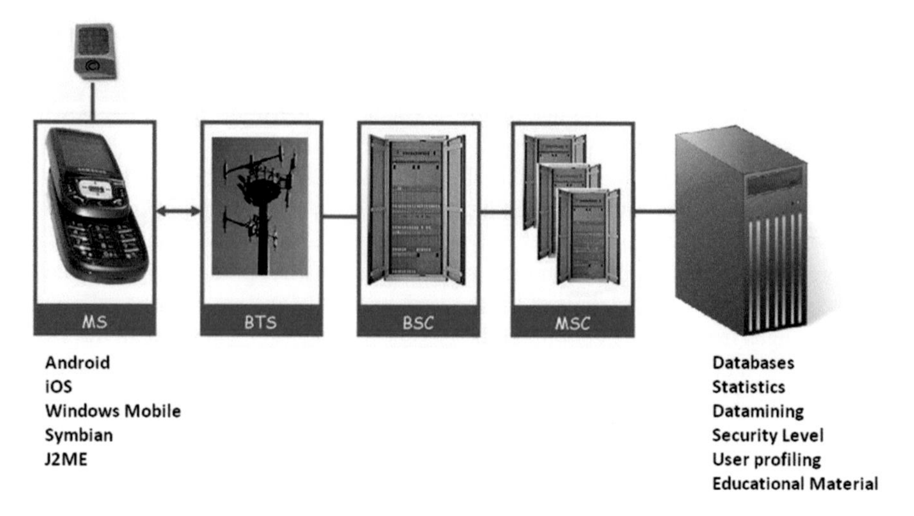

Fig. 5.25 System architecture

Chapter 6
An Example of a Malware that Can Be Used for Industrial Espionage or as a Personal Spyware and a Way to Protect from It

6.1 Introduction

As we have mentioned in previous chapters, there is a great number of software that could be installed on a smartphone for the purposes of industrial espionage. In this chapter we will be examining how a new technology could be used in malware targeting Android devices which in turn could be used to infect an Android device and intercept confidential information as well as a way to protect against that new breed of malware. The chapter is based on the paper "Fragkiskos-Emmanouil Kioupakis and Dr. Emmanouil Serrelis, Defending Against Android Malware that Uses Tor as Their Covert Communications Medium," International Journal of Cyber-Security and Digital Forensics (IJCSDF) Vol. 4 No. 4: 469–481."

The consumer base of devices based on Android OS is rapidly increasing. The number of consumers using such devices for security critical tasks, such as online banking transactions and accessing corporate networks, is also increasing at a similar rate, making them valuable targets for malware practices. To date, even though significant or large-scale attacks involving Android devices have failed, security is proven to be a critical aspect for both individuals and corporate users. The type of usage of those consumers ranges from day-to-day communications to accessing sensitive corporate data. Especially when it comes to corporate users, Android devices are often used for storing critical information such as business contacts, sending sensitive information via email, visiting corporate network resources, and sending SMSs to business partners.

The insufficient security management from the users' side in combination with the lack of security within the relevant applications (e.g., implementing transaction authentication numbers (TANs) for online banking transactions) makes those devices a valuable target for malicious users regardless of their motives. Up until recently, the lack of attack vectors that contribute to the distribution of infections and the low sophistication of malware applications have allowed the Android OS model to avert significant attacks from malware.

I. Androulidakis, F.-E. Kioupakis, *Industrial Espionage and Technical Surveillance Counter Measurers*, DOI 10.1007/978-3-319-28666-2_6

6.2 The Tor Network

Initially developed as a project of the US Navy for intelligence gathering, Tor is a network of relay nodes used to form random virtual tunnels for a computer to use in order to reach websites or other Internet services. Instead of directly connecting to an online resource, using Tor as a connection is made to an entry node which encrypts the contents and passes it to an intermediate node which adds another encryption layer and forwards it to the exit node. Finally the exit node connects to the desired destination.

After that process the traffic follows the reverse path to reach the user. Each node uses a different layer of encryption, thus the name "The Onion Router." That architecture allows the entry node to be aware of who is making the request and the exit node to be aware of the requested website. This information combined could reveal the whole path so the intermediate node is used to prevent that, and also in each step different encryption layers are used. Furthermore, the Tor network could be used in a reversed way, as in the website itself using the Tor network to hide its identity and location. Websites that follow this discipline have an address ending in .onion, and they are generally accessible only through Tor.

Tor is being used by individuals that do not want to be tracked by websites or to connect to websites and services that their ISP blocks access to in certain occasions of censorship practices. Journalists use Tor to communicate with whistleblowers or to publish articles online without the fear of being arrested while being in countries with authoritarian regimes. Besides protecting the freedom of speech, the technology named Tor has also given the ability to malicious users to proceed to various acts—mostly illegal—with various motives. Recently it was used as a component of malware applications, creating in that manner a new category of threats that requires a new perspective of defense mechanisms for detecting and mitigating those new threats.

6.3 Current Threat Landscape

To date, there have been only two confirmed cases of malware that use Tor as their covert communication channel component. In the first case, the malware called "Simplocker" is a ransomware type of malware that encrypts certain type of data located on the smartphone device. The difference between this particular ransomware and others is that "Simplocker" uses a command and control (C&C) server operating on an .onion domain, and Tor network is used to establish communication with it. The purpose of this communication is for the malware to acquire a list of file name extensions, in order to encrypt the corresponding files on the device. However, this implementation does not exploit the full potential of Tor as a covert communication channel, limiting the potential impact of the malware. With minor changes in the malware payload code, the malware could potentially upload the device's files to the C&C server and then delete them from the device, making the recovery of the files much harder, either upon infection or later as a means of escalating the pressure against the victim and in order to pay the ransom.

In the second case, the malware named Backdoor.AndroidOS.Torec.a is reported to be the first Android Trojan to use Tor. This particular malware constitutes the "Orbot" Tor client repackaged with a malicious payload. Upon infection, the malware establishes communication with a C&C server that operates on an .onion domain of Tor network. The purpose of this communication is to transfer data collected as JavaScript Object Notation (JSON) objects from the Android device to the C&C server. These data include the phone number, the country, the IMEI number, the phone model, and the OS version. The malware also includes a function used for intercepting SMSs, sending data regarding Unstructured Supplementary Service Data (USSD) requests to the C&C server, as well as to send a list of all installed applications on the device. The intercepting SMS function is also used to issue commands from the C&C server to the infected device.

Another case of malware that uses encryption to conduct its malicious activities is the malware identified as "Red October." Although its main target was PCs, it was able to extract data from mobile devices that were connected to the infected PC. "Red October" was used to conduct cyber espionage against embassies and government organizations but also research institutes and enterprises. Similarly to other cases, it used an encrypted communication channel to contact C&C servers. The usage of encryption that is only compatible with PCs limited the malware's capabilities to conduct mobile-specific malicious activities since it required the mobile devices to be connected to an infected PC.

One more example of relevant malware is the one identified as "OnionDuke." This malware used a malicious Tor exit node to wrap legitimate executable files requested by the user with malicious code. By doing so, the original legitimate file was downloaded along with the malicious code limiting the possibility of discovery by the user and increasing the chances of bypassing integrity check mechanisms. This practice resembles the repackaging method of distributing malware as mentioned above. Upon infection, the malware contacted C&C servers to receive orders or download additional malicious code. It is also believed that the same malware was deployed in the form of a targeted advanced persistent threat (APT) against government agencies in Europe. Even though this malware only targeted PCs, it demonstrates another way of exploiting the Tor network for malicious activities.

The combination of technologies and practices mentioned above along with the implementation of Tor could lead to the creation of a more sophisticated malware with extended capabilities that could be used for cyber espionage by infecting mobile devices of specific individuals such as high-level executives or other people in the organization who have access to sensitive corporate information. To date no commercially available detection mechanisms have been found that are capable of detecting the initiation of Tor connections at its source (i.e., at the Android device level). Typical mechanisms (e.g., firewalls) found in most enterprise environments are only capable of explicitly forbidding Tor traffic from passing through them. Executives using their smartphones outside of the company or not connected to the enterprise network when they are on premises are still fully exposed.

6.4 What Would Make Such a Malware Successful?

Initially, since the infection requires physical access to the device, the time frame required for the infection is considered to be critical. This time frame should be at most equal to the average time that users leave their devices unattended. For the malware described here, the time targeted is less than one minute which is empirically considered less than the time mentioned above. During that time, the attacker will be connected through https to the web server that hosts the malware and install it. To speed that process up, the host's URL can be written on an NFC tag that can be used with NFC-enabled devices in order to avoid the manual access to the device's browser and the typing of the URL.

Upon infection, the victim should not be in the position to detect the operation of the malware. For this reason, the malware operations should not interfere with the rest of the device's operations when the victim uses the device. For example, when navigating to a menu, there should not be any delays or glitches or any other behavior that might make the user suspicious. Hence, while the user browses through the contact list or the call log, there should not be any abnormal activity such as time delays or screen movement glitches, even if the malware is intercepting data at the same time.

For the same reasons, the malware should not appear in the list of installed applications nor in the list of running applications. These two requirements can only be implemented through OS-based exploits and cannot be based on permission exploitation like most of the requirements. Such exploits may differ from an OS version to another. In any case, the malware should be compatible with the most widely used Android version.

The malware should be sending the extracted data to the C&C server in specified time intervals, when the victim's smartphone device is connected to the Internet. Though, to ensure the integrity of the malware operation, the extracted data should be sent to the C&C server after a predetermined time threshold regardless of whether the victim's device is connected to the Internet or not. The data to be extracted by the malware include victim's geolocation, call log, contact list, SMSs list, and browser bookmarks and history.

As far as the network traffic is concerned, the malware should be using the Tor anonymity network in order to secretly exchange data and control commands between the infected device and the C&C server. In addition, the tracing of the C&C server that the malware uses should be made very hard due to the use of Tor. Furthermore, the malware should easily operate in enterprise environments dealing with advanced network blocking mechanisms that enterprise security solutions have implemented.

For that reason, the Tor traffic of the malware should not be different from any other Tor traffic, hence making it difficult for anti-malware mechanisms to block malware Tor traffic while letting regular Tor traffic to pass through. The malware should include controls to determine when is the most appropriate time to create a covert connection to the C&C server, based on specific events. An example of such event is when the device is in idle mode (e.g., the screen is turned off). This would also achieve the enhancement of the covertness of the operation. For the same reason, when the malware is active, it should not trigger any notifications to the user.

To further enhance the robustness of the malware operations, there should be anti-tampering mechanisms implemented, in order to make more difficult for the security analysts to manipulate the data entered to the malware. More specifically, this requirement aims to ensure that no false data are fed to the malware ensuring that the attacker receives only useful data from the device in case of detection.

6.5 Technical Description of the Malware Operation

For designing each subsystem (module) and operation of the malware, an Architectural diagram has been made and is presented in Fig. 6.1. The malware's operation is highly dependent on proper timing. In order to achieve that, it uses the system time in combination with an internal counter. During malware initialization the internal counter is set to zero. Initialization could occur either upon infection or upon turning on the device after infection. Ten minutes after initialization—calculated by the difference between the system time and the internal counter—the malware checks if the victim is on the move. If that condition is true, then the counter resets, and the process proceeds to the intercept module. This process is performed by the malware in order to intercept the victim's geographical location (geolocation) along with the changes made to the call log, the contacts list, SMS list, and browser history and bookmarks.

If the application exits the initial ten-minute loop, this indicates that the victim is either not moving or he/she was moving and now has stopped. In this case, at a thirty-minute mark, indicated again by the internal counter, the application enters a second loop which checks whether the victim is using the device or not. If that condition is false, the application proceeds to the intercept module and resets the counter so the process begins again. If that condition is true, which indicates that the victim is using the device at the thirty-minute mark, the application enters a secondary loop which also includes a secondary internal counter. That secondary internal counter awaits for another thirty minutes for the victim to stop using the device in order to proceed to the intercept module.

If the victim continues to use the device for the thirty minutes counted by the secondary internal counter, the application exits that loop and enters the last loop which at the sixty-minute mark resets the primary internal counter and proceeds to the intercept module with no further checks. This process aims to send the extracted data to the attacker at most within an hour regardless of conditions ensuring that these will be ultimately communicated no matter the usage circumstances. This requirement is related to the robustness of the malware. Following any of the above cases, the application then proceeds to the intercept module, seizing the geolocation data, the call log, the contacts list, the SMSs list, as well as the browser history and bookmarks.

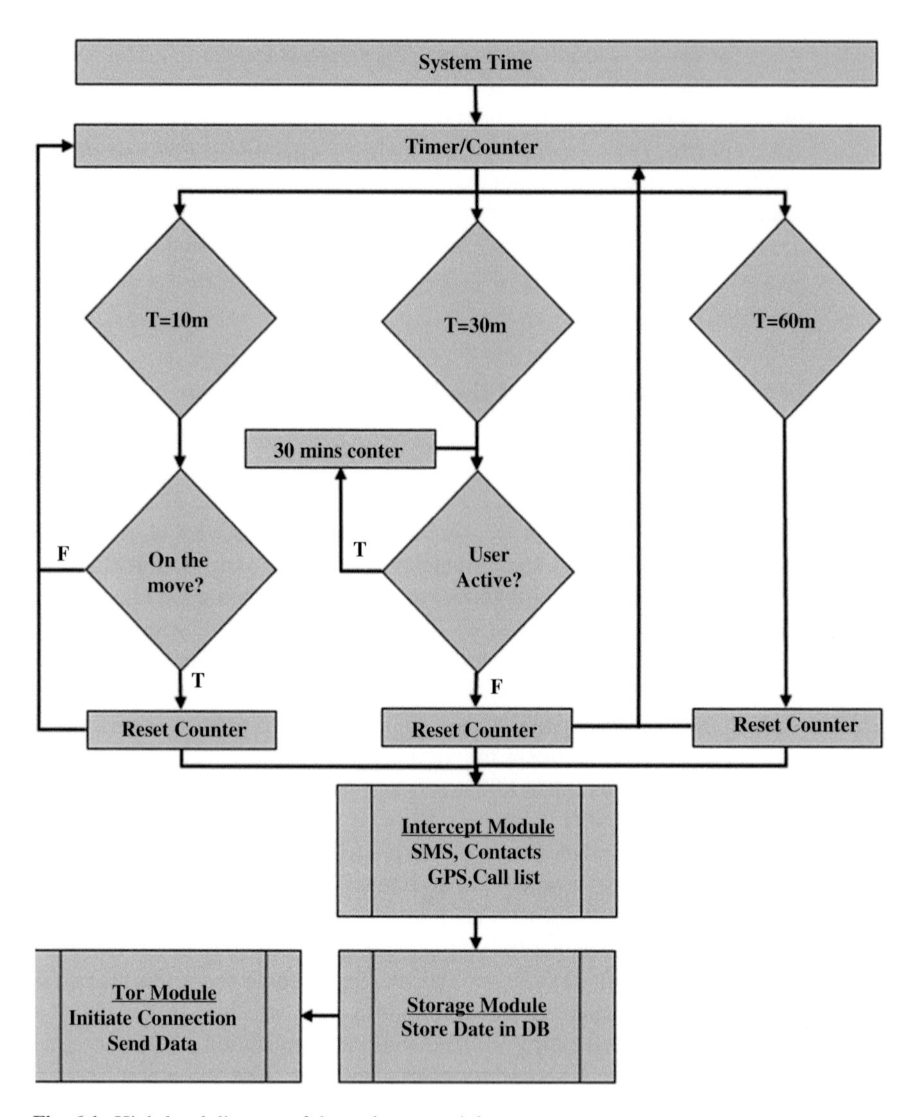

Fig. 6.1 High-level diagram of the malware modules

Following, the data are received by the storing subsystem (storage module) which handles the storage of the intercepted data to the database included in the Android OS. This process is implemented so that all data can be manipulated as a single entity. Also when storing the data, a time stamp will also be included (according to the system time) in order for the attacker to be aware of the exact time of each data intercept cycle. In this manner, the attacker would also be aware if any of the data intercept cycles was delayed, skipped, or otherwise altered. Subsequently, the entity created by the storage module will be forwarded to the Tor subsystem (Tor module) which in turn is responsible for the connection initiation and session

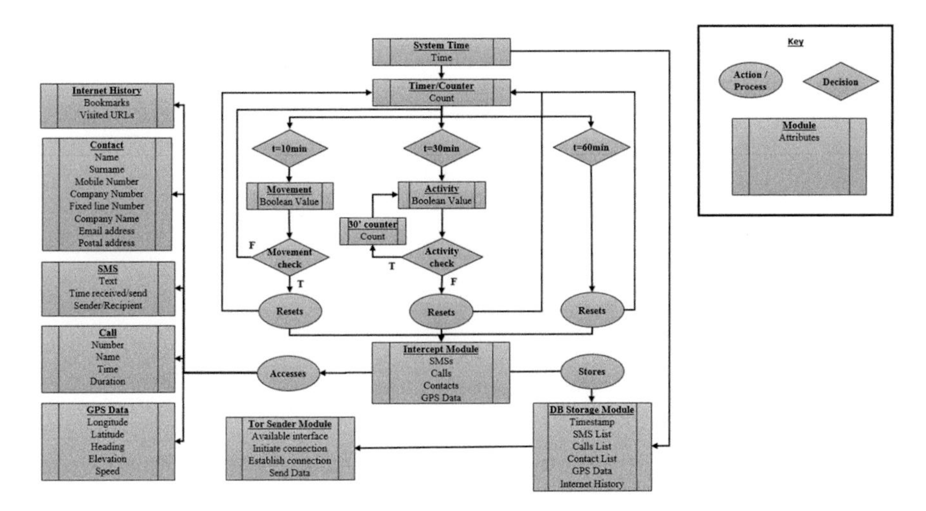

Fig. 6.2 Architectural diagram of the malware modules

establishment with the C&C server. Upon that, the intercepted data will be sent to the C&C server, and that completes that data interception cycle.

6.6 What Would Make Countermeasures Effective?

Various static and dynamic detection methods can be used to detect the malware behavior of such malicious applications as shown in Fig. 6.3.

Of these, static and permission-based analyses are not appropriate due to the offline nature of their implementations. Since a real-time detection implementation is required, dynamic analysis is the only appropriate method for detecting such a malware. The suggested detection methods can be classified, according to Fig. 6.2, as network traffic analysis, since it detects connections with specific destination ports as described in the corresponding section below. Moreover, a different detection approach is introduced, which includes the detection of a string, transmitted during a Tor connection initiation process which cannot be classified as pure network traffic analysis methodology.

To add to this, a native code analysis would fail to detect such a malware since the core code itself is not malicious and since the malicious processes are typically based on permission exploitation. However, regarding offline implementations, other approaches could be followed upon detection and extraction of the malware. Static code analysis would reveal to an experienced researcher the true nature of the application and the way it intercepts and sends sensitive personal data. Moreover, permission-based analysis would reveal the exploitation of permissions which leads to the malicious behavior of the application.

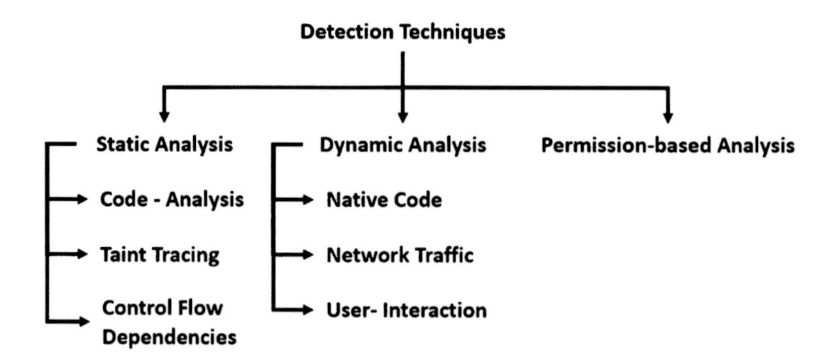

Fig. 6.3 Malware detection techniques categorized by type

One of the most critical parts of the anti-malware is the one that will detect the initiation process of a Tor connection. Upon detection of such activity, a decision-making subsystem will determine whether the application requesting the Tor connection would be granted permission to establish this connection or not. The decision-making process takes into account whether the Android process requesting the initiation of a Tor connection is generated by a signed application. If this is not the case, no permission would be granted, and the connection would be automatically blocked.

Still, even when the application is signed, the anti-malware will prompt the user whether he/she permits the application to initiate a Tor connection or not. The anti-malware could include a user and/or vendor customizable application whitelist that should include all applications authorized to use Tor, in order not to block unsigned applications that the vendor or user trusts and not to prompt the user for signed applications that have been authorized in the past by the user.

The anti-malware operation should be transparent to the user. Only notifications regarding the blocking of an application that tries to initiate a Tor connection and the prompt for asking permission for signed applications should be generated. Limiting the notifications toward the user would minimize user annoyance caused by unnecessary information while also making more efficient usage of the system resources. In fact, the only time that this anti-malware would be taking any action would be when detecting a Tor connection being initiated.

For further enhancing the efficient usage of system resources, the anti-malware should be in a constant idle-standby state and exiting that only when taking action is required (i.e., a Tor connection initiation by an application). The scanning subsystem should remain active even when the device is in an idle state (i.e., screen turned off), since a malware could be initiating the Tor connection at exactly that state to increase its covert nature.

The anti-malware should aim not to exhaust the battery or make heavy use of system resources by deactivating specific modules based on various events. An example of this includes deactivating all components when the device is in "flight mode." Subsequently, the module for traffic scanning could be divided in separate sub-modules equal to the number of different interfaces found in various devices in order to deactivate a certain sub-module when the corresponding interface on the device is also deactivated.

Furthermore, the anti-malware operation should not cause any conflicts with other anti-malware software running on the same device and should be compatible with as many Android OS versions as possible, starting with the most widely used version. Moreover, the anti-malware could come in two different versions, one that would be running constantly as a background service, providing real-time protection, and a standalone version that would be used on demand, scanning running applications when the user suspects that there is a Tor-enabled malware application installed on his/her device.

Finally, as far as performance and reliability is concerned, the anti-malware should not give a false alarm rate of more than 5 % while not failing to detect Tor connection initiations in a rate greater than 5 % as well. These percentages are close to the average industry performance, and they are a direct measure of reliability.

6.7 Technical Description of the Anti-malware Operation

The anti-malware implements a listener/handler module which communicates with the network layer which in turn communicates with the physical interfaces that are present on the device. The listener/handler module acts only when it monitors the traffic and detects any Tor traffic being initiated on one of the interfaces. For the decision-making subsystem, the Tor connection detection is accomplished by detecting traffic using destination ports 9001–9004, 9030– 9033, and 9100 along with a string that is transmitted during Tor connection initiation.

Upon detection, the module checks which application is requesting to initiate the Tor connection and checks whether that application is included in the application whitelist. During inspection, if the application is indeed included in the whitelist, the listener/handler module takes no further action. In case the application requesting to initiate the Tor connection is not in the whitelist, then the anti-malware application proceeds to the next inspection step. Any application that requests to initiate a Tor connection that is not included in the safe list of applications authorized by the user will be inspected to determine whether that application is signed or not. In case the application is indeed signed, the anti-malware application will prompt the user asking to authorize that application to initiate a Tor connection.

If the user authorizes that application, then it will be added to the safe list of applications in order not to prompt the user again in the future for the same application, and the anti-malware application will take no further action, allowing that application to proceed with initiating the Tor connection. In case that the application requesting to initiate the Tor connection is either not signed or the user does not authorize it when prompted, the anti-malware application will deactivate the interface that would be used for the Tor connection in order of the user to investigate that application and uninstall it if needed (Fig. 6.4).

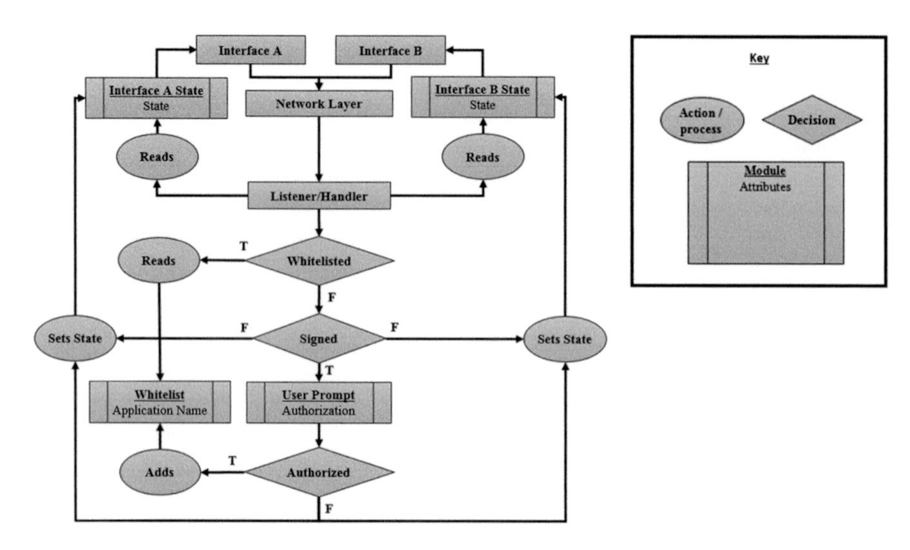

Fig. 6.4 Architectural diagram of the anti-malware modules

Furthermore, the anti-malware listener/handler module will be constantly monitoring the state of each interface and deactivate the corresponding scanning submodule if an interface is deactivated in order to make use of system resources more efficiently.

6.8 Conclusions

The threat that Tor as a technology poses is existent and in many cases underrated. Appreciating the fact that today's information translates into power and competitive advantage, the mitigation of new threats and the development of appropriate and sufficient defense mechanisms are considered to be critical. With companies counting losses of hundreds of billions of dollars annually due to cyber espionage, any new threat that could be used in that manner should be seriously taken into account. If deployed in the APT form, the malware could be used for corporate, government, or industrial espionage by infecting devices of high-ranking officials or high-level executives, respectively. For enhancing the propagation/infection methods, various exploits for remote exploitation can also be used. In all cases, once a device is infected, it should be able to operate for a significant amount of time before being detected, especially when it comes to average-skilled users. As users tend to store more and more of their private data on a variety of devices, the necessity of protecting privacy is more critical than ever.

All these facts signify a new trend for malware threatening mobile devices. The nature of open-source software allows and encourages the creation of new software products, but likewise new threats can arise using the same techniques and technologies. Therefore, it is imperative that information security professionals need to mitigate sufficiently. The very same nature of open-source software that can pose threats can also be proven a valuable ally against the challenges that arise when creating defense mechanisms that mitigate threats that threaten private and enterprise environments.

Chapter 7
Protection Against Industrial Espionage

7.1 Introduction

In the last chapter of the book, we will be discussing the aspect of protection against industrial espionage and counter-espionage methods both on the technical and the nontechnical level. At the same time, we will be examining the protection options that patents and trade secret safeguarding can offer. The book concludes with presenting the methodology for detecting interceptions along with guidelines for selecting the appropriate contractors for that process.

7.2 About Protection in General

The challenges and problems that the modern businesses face are bigger than ever. The market environment is extremely competitive. Products need to be placed on the market without any delays, and information needs to be transmitted as quickly as possible, often with sacrificing security. Besides, the more innovative a product is, the greater the danger is for its design to be copied. At the same time, new technologies introduce new vulnerabilities in the systems. It is easy to understand that protecting the intellectual property is vital not only to the competitive advantage but to the company's existence as well. Alongside that, in the ethics side, the copyright of the creators and inventors must also be protected.

Protection is a multilayer process aiming to perverse the confidentiality, integrity, and availability of information. It includes legal aspects (patents, trade secrets), physical and IT security (equipment, software, and hardware), IT governance (policies, roles, responsibilities, policy enforcement, auditing mechanisms), financial restraints (budget), and of course training.

Choosing the best strategy for protecting intellectual property depends on a number of factors: the available alternatives, the kind and degree or protection they

I. Androulidakis, F.–E. Kioupakis, *Industrial Espionage and Technical Surveillance Counter Measurers*, DOI 10.1007/978-3-319-28666-2_7

offer, and the cost and eventually the value that this protection can have for its owner. In any case, protection and security, for both the information and the infrastructure itself as well as for products or services, require procedures that are quite demanding on resources. Human resources, appropriate technologies, and policies for IT governance are needed, while at the same time, this overhead should not be affecting everyday operations. Of course, complicated solutions worsen the situation, while in an era that information is omnipresent, traditional methods and conceptions regarding security are no longer effective.

The options for managing intellectual property assets that we will be examining in the legal domain are two: publicly registering the information in order to protect it (e.g., submitting a patent) or alternatively keeping information confidential.

7.3 Patents

The first thing that comes into mind for protecting business advantages is acquiring a patent. A patent is a title of protection that is provided to the beneficiary for new inventions containing an inventive step over the state of the art that are also able to be exploited in an industrial way. An invention is judged as new if it surpasses the state of the art and includes an innovative step, i.e., if according to the experts the invention is sufficiently inventive (in a nonobvious way) compared to the current state of the art. The invention is considered to have industrial applicability as long it can be produced and used in any sector of the industrial production.

A patent grants the owner the right to exclusively use, sale, and make use of the invention in general, excluding third parties from the provisional rights of the "claims" of the patent. The claims are the elements-characteristics-methodology that the applicant is requesting protection for. Attention should be given to state those properties correctly, thus adequately protecting the invention; otherwise, it is possible to "open the competitors' eyes" without succeeding in protecting the intellectual property. If the patent is not written properly, it can describe the idea of the invention so that it can be copied, but the claims are not formatted probably, so a competitor can effectively implement the idea without infringing the patent.

By having a patent, a company can enjoy a business advantage in the market since it forbids its competitors from making use of its idea/product. That way, confidentiality of information is (legally) protected. Not because it is confidential but because the company took the time to register it on its name. So, in case some other company steals its idea, the company owning the patent has a powerful legal "weapon" that can be used to protect its rights in accordance with the law.

The main advantage of owning a patent is that the company could then promote the product in the market without the fear of it being copied (but again, the company should be in the position to discover such attempts and legally prosecute the ones responsible for those attempts, which is not that easy if those attempts are taking place in another country or even in another continent). This legal protection in fact can be direct, since it is not necessary to prove the intellectual property theft as long as the competitive product is infringing the claims of the patent. So the patent owner does not have to prove that an actual incident of intellectual property theft has taken place. Respectively, the company does not have to worry anymore about high-level executives or engineers moving to other companies, bringing along the inevitable expertise leakage, since the product would be already protected by the patent.

A simple process for obtaining a patent includes:

- Submitting the application
- A deadline for possible additions or corrections
- Examination of the innovative aspect of the invention and its usability yielding a research report
- A deadline for the applicant to submit comments on the research report
- A final research report
- Issuing the patent

An interesting detail is that the invention should not have been previously published or presented in any way to the public before submitting the application for the patent. So, a member of the academia who will be publishing his research in a scientific journal would be preferable to first fill a patent application of the inventive part of his research.

It is often the case that patent rights are sold (or, often, buying out of entire companies takes place just in order to gain access to certain patents that this company holded). Small companies following that strategy have brought huge profits to their owners. At this point of course, there have been malicious practices by companies that rush to patent even the smallest element, expecting to sue other companies that inevitably will be using that element (while there is always the possibility that someone else could get to the same invention with his own research without any fraudulent practices). Imagine what would happen if the typical "qwerty" keyboard was patented. Then every keyboard today would have a different layout!

Especially in the ICT industry, "patent wars" is a very common practice, and the readers surely have heard about conflicts between industry giants (e.g., in the mobile telephony manufacturers). This phenomenon isn't new. Even the invention of the telephone itself hides a very interesting story behind it, with conflicts between Bell and Gray regarding patents.

7.4 Confidential Information, Know-How, and Trade Secrets

A patent isn't always the best way of protection. The main drawback of that process is that it requires publishing and adequately describing the invention/idea/methodology and actually in a way that it can be successfully implemented by the appropriate expert. Sometimes though, there is information that has to be kept secret.

Information (e.g., blueprints, recipes, procedures, methodologies, diagrams, pricing policies, strategies, etc.) used by a company for its commercial activities that are not revealed to the public are known as "confidential information" or "trade secrets." The most typical example in this case is the secret recipe for the production of the most well-known and consumed colas.

These secrets can either be in a physical or virtual form, stored in any way, physical or electronic. In order for information to be considered as confidential, it needs to be secret and have an economic value (by providing a competitive advantage), and the owner needs to have taken the appropriate measures to keep it secret.

The main advantage here is that it can be applied to anything (for as long as the secret is not revealed), and it does not require the expensive and time-consuming processes of filling a patent. Of course no publication at all takes place. On the other hand, the disadvantage is that there is always the possibility that someone else legitimately (by pure chance or actual research) gets to produce the same or a similar product without intercepting that information and possibly be better protected by actually filling for a patent.

Since the value of such information solely depends on the confidentiality, it is imperative to minimize the exposure of the information and to limit the number of people that have access to it. In case it is needed to transmit this information to third parties, then this should always happen with a nondisclosure agreement between the involved parties.

The methodology for protection through trade secrets includes identifying the trade secrets, recording them, evaluating their value, and categorizing them according to the threats and vulnerabilities that identify the potential risks. Protecting them involves, among other things, an official IT governance plan, nondisclosure agreements, marking, tracking, secure storage, training, and finally legal action plan in case of leakage.

Especially regarding training, its importance can never be emphasized enough. If employees do not know what exactly they need to protect them, how are they going to do it? Personnel that produces and manages commercially valuable information needs to know its responsibilities in regard to the trade secrets. Raising awareness concerning the threats of industrial espionage, intellectual property theft, piracy, and copying is needed as well. Finally, we should not forget the importance of training-informing contractors and collaborating parties as well.

If confidential information gets published without prior authorization, it is possible to take legal action. In this case, it is important to be able to prove that the information leaked was indeed confidential and all the necessary precautions (to a reasonable extent) for its protection had been taken. The auditing mechanisms for the protection of information would verify that the incident took place and will collect and preserve the necessary evidence and traces so that law enforcement could be led to the perpetrators.

7.5 Equipment for Detection of Electronic Tapping

Having analyzed both the background of industrial espionage, as well as the intellectual property assurance measures at a legal protection level, we will continue with the equipment used to detect eavesdropping devices and then describe its exact methodology.

Fig. 7.1 Frequency receiver

7.5.1 *Frequency Receivers and Scanners*

One of the main detection tools is the frequency receiver (Fig. 7.1), which is basically a special radio, i.e., it receives radio frequencies. A good frequency receiver must have very high-sensitivity and excellent frequency resolution (the ability to differentiate among emissions in close by frequencies) along with the possibility of rejecting unwanted signals located near the signal to be tested. It should also be able to scan a large part of the radio spectrum. It is important that the receiver should be able to demodulate various modulation techniques (AM, NFM, WFM, SSB), so that the operator will be able to find exactly what is emitted (in analog broadcasting). Scanning for wireless cameras is performed with the equivalent equipment (Fig. 7.2) by demodulating programming broadcasts of AM-TV and FM-TV type. The use of the appropriate antenna for each frequency band under analysis is particularly important. Most models include rapid automatic band scanning, stopping each time they detect an emission; in this case, they are called scanners.

Even more sophisticated scanner models, such as the one in Fig. 7.3, are computer controlled and rapidly sweep the range, so that they can record even very short emissions, approaching in functionality the spectrum analyzers, which will be described below.

In any case, the operator will scan the wider possible range and will examine all suspicious signals, excluding emissions that may originate from radio stations, radio-taxi services, security forces, cordless phones, wireless networks, and any other legitimate source. The process is very tedious and time consuming, because,

Fig. 7.2 Frequency receiver capable of receiving television signals

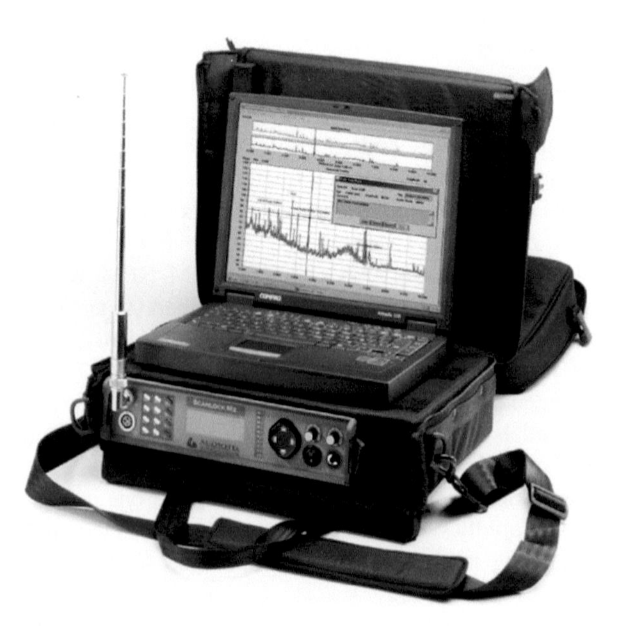

Fig. 7.3 Sophisticated scanner

depending on the area, one can identify hundreds of signals that must be sorted and examined to exclude the possibility of an eavesdropping device.

7.5.2 Spectrum Analyzers

At a higher cost, the spectrum analyzer is a tuning receiver capable of scanning large frequency bands, providing a visual display of the power of the emissions in each radio frequency it scans (Figs. 7.4 and 7.5). The horizontal axis on its screen depicts the signal frequency, and the vertical axis shows its power, thereby reflecting the emission background in this region. By analytically depicting the radioactivity in the spectrum, it helps the operator to select at a glance the areas which he will

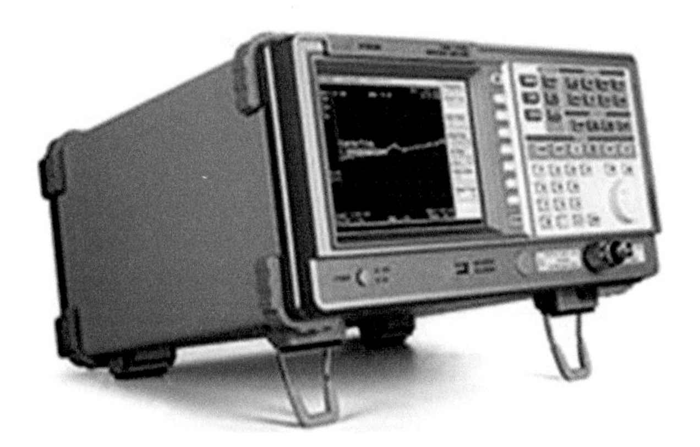

Fig. 7.4 Desktop spectrum analyzer

Fig. 7.5 Portable spectrum analyzer

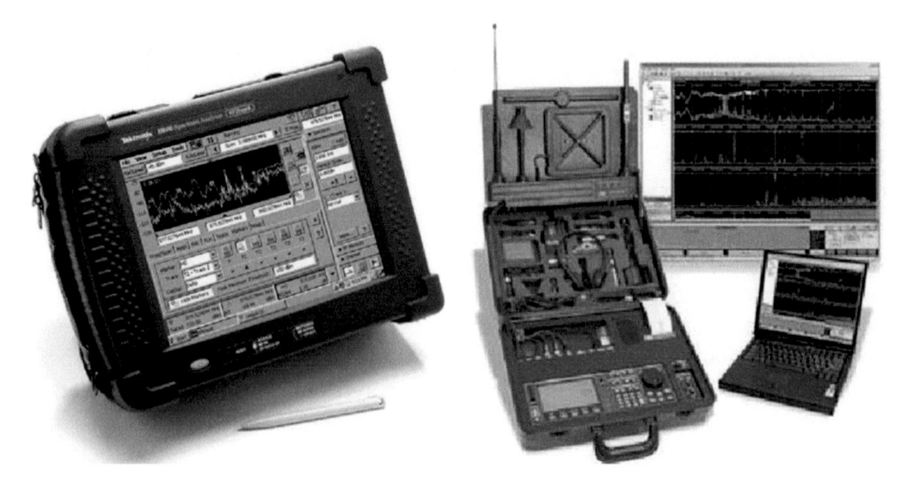

Fig. 7.6 Special spectrum analyzers-correlators for tapping detection

then further analyze more carefully. Obviously, the operator can adjust the central frequency, the sensitivity, and the detection range span. Besides, the analyzer may also have a demodulation circuit; so the broadcasted signal cannot only be viewed but also be heard. Depending on the waveforms shown, the skilled operator is able to discern the type of modulation detected by the transmitting device. Indeed, a modulation in a carrier will cause its image on the analyzer to change in a predictable manner. Its main advantage over a simple receiver is just the fact that the visual examination can detect modulations that a simple receiver is not able to demodulate (e.g., digital modulation), as well as techniques such as frequency hopping. A key point to be noted is that the analyzer must have sufficient sensitivity to detect devices that emit in very low power; otherwise, it will not be able to "hear" them.

Especially developed, military-grade spectrum analyzers (Fig. 7.6) along with the appropriate software can collect, classify, analyze, correlate, and identify (using radio direction finders) radio sources almost automatically. They can also separate emissions that take place on the same frequency, e.g., the weak signal of a malicious device which is masked by a television signal. Furthermore, they can also record a signal in a given areas (background emissions) for several hours and later detect the presence of new and potentially suspicious emissions. Meanwhile, they can demodulate dozens of modulation standards, both analog and digital. As expected, their cost is equivalent to their potential, reaching up to 100,000 euros.

7.5.3 Broadband Receivers

The wideband receivers (broadband receivers) rapidly sweep the entire radio spectrum, calculating the average energy contained in it, without distinguishing individual frequencies. In that way, they can present a direct indication of radio

Fig. 7.7 Broadband receiver

emissions. Some models, like the one on Fig. 7.7, have an elementary demodulation circuit so that the operator can listen to the possible emission. A key problem of this equipment is that stronger emissions from neighboring legitimate transmitters can overmask a low-power suspicious emission. Their usefulness is also limited in terms of digital broadcasting, as they cannot demodulate them.

Broadband receivers usually include the audio feedback technique. If an analog emission device is located when scanning, then the receiver receives the audio, strengthens it, and, as this process is continuously repeated, it causes feedback which leads to possible loud wheezing. The reader may have noticed the similar procedure if he turns a microphone toward a speaker. This technique is usually avoided because the loud sound will warn the spy that his bug was found and he will react accordingly.

Widely advertised cheap radio-frequency detection devices (Fig. 7.8) constitute a sort of broadband receiver, but in no way approach the usefulness of professional equipment. Their mode of operation is based on the detection of radio frequencies and electromagnetic fields using a diode which receives an extended radio-frequency range and then amplifies it and displays it, visually with a LED indication or with an audible hum. Whenever they receive a signal, the operator lowers the sensitivity and rescans the region in which it was identified. By decreasing the sensitivity, the receiver must now approach very close to the transmitter in order to detect it. Therefore, with successive scans, it eventually reaches the full reduction of sensitivity level, and along with the removal of the antenna, the receiver reveals the exact emission point, now located a few centimeters away.

Unfortunately, their simplistic electronic design does not allow for good resolution of frequencies; thus, in an area with strong electromagnetic activity, like in all modern cities, they literally go crazy by receiving signals from every corner of the test area. Respectively, their low sensitivity does not allow them to discriminate a suspicious device whose signal is covered by another, more powerful, legitimate signal (e.g., radio). Furthermore, widely advertised models require extra attention,

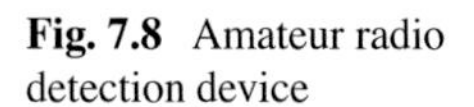
Fig. 7.8 Amateur radio detection device

because their purchase demonstrations and advertisings include various deceptive techniques, e.g., the emission device will transmit its signal in the exact frequency where the receiver has a better sensitivity, and thus, the results look surprising, whereas in practice, they are less than moderate.

7.5.4 Field Meters

The field meter has the same operating principle, i.e., the scanning of the whole spectrum at once, but in a more evolved fashion. Not only it identifies radio-frequency sources but also measures their power, which is particularly useful for potentially health-hazardous emissions. They generally have better electronic circuits and may switch among different operating frequencies, but again their main disadvantage is the lack of tuning in specific frequencies. A key feature of their operation is the very accurate power measurement, which raises the cost without offering something substantial in the counter-surveillance scan work.

7.5.5 NLJD: Nonlinear Junction Detectors

When it comes to eavesdropping detection, most people have in mind the particular equipment shown in Fig. 7.9. This heavy machine (that looks like a gold detector) has a very significant advantage over others. It can detect any electronic eavesdropping device (containing diodes, transistors, integrated circuits, condenser microphones, and semiconductors in general), even if it does not emit, has ceased to operate, or is not powered at the moment of detection! This is possible because it

Fig. 7.9 Detectors of nonlinear contacts

does not detect active emissions but rather detects the existence of solid-state junctions (p–n). In a nutshell, the device detects the building components of electronic devices.

Technically, the detector is a microwave frequency transceiver. Its head consists of an antenna (or a combination of two antennas) that transmit and receive microwave radiation. The receiving circuit is not tuned to the broadcast frequency but to its harmonics (harmonic frequencies are the integer multiples of the fundamental frequency of a signal). The operating principle is based on the observation that a semiconducting device containing a different type of semiconductor junctions (p–n) tunes in in the presence of suitable radio-frequency signals (in the microwave band) and further transmits in the second and third harmonic frequency of the signal it has received. Therefore, the detection of harmonics indicates the existence of such a device at a certain distance from the detector head, inside walls, ceilings, and other objects. The operator is informed by monitoring visual signals or by an audio signal through headphones.

As with the rest of the equipment, this machine has some drawbacks. Coming close to certain materials, e.g., different metals-wires, can give a false indication. The power output is also proportional to the detection rate. Shielded devices (such as microphones) may escape detection. The increase in power is not a viable option, as there are limits both in terms of avoiding radio-frequency interference, as well as for health reasons.

At this point we can mention the well-known example of the US Embassy in Moscow during the Cold War, since it is a typical example of how an NLJD can become useless in (specific cases) bug sweeping. During the construction of the building, apart from actual interception devices, the cement used was further deliberately mixed with thousands of diodes, not as part of functional electronic devices but as simple components. In that way, any future detection attempt using nonlinear array detectors would receive false indications everywhere (detecting the implanted diodes), thus making the separation and identification of a truly malicious device near to impossible. The full story is very interesting, that's why we will spend some more paragraphs discussing it.

In 1969, the US government signed an agreement with the Soviet Union to provide an embassy building in Washington in exchange for a new embassy building in Moscow. The agreement stated that the Soviets would have control of the design and construction of the embassy building in Moscow.

Since the USA had the right to do all the finishing work such as inside walls, windows, and doors, there was little concern that the Soviets would be able to implant bugs that could not be detected. That assumption was made also because the USA had security personnel dispatched at the construction site with the purpose of detecting possible eavesdropping devices which was thought to be a sufficient measure.

The Soviets were building precast concrete pieces for the embassy building at Soviet factories away from the building site, thus out of the view of the American security experts. Due to several construction delays, even by 1979 that the precast concrete pieces were arriving at the construction site, there was still no clean plan for the American security experts regarding what were supposed to look for or the on-site security needs.

Even when suspicions arose from the inspections at the construction site, they weren't listened to since the US government was rushing to get the job done under the pressure of the Congress because of cost overruns and poor results regarding similar constructions in other capitals. Instead, the US government focused on the construction being inferior to the American standards, and the project was being treated as a cost-driven operation. At the same time, the US government still believed that they would be able to remove any eavesdropping equipment that might be found at any time and could "clean" the building even after the building was complete.

As construction of the outer structure was close to completion, the Soviets rushed to complete the top floors of the building where secret embassy operations would be taking place. At the same time, an external Soviet-operated freight elevator was mysteriously disabled, meaning that workers needed more access to the top floors.

It took till early 1982 for a specially trained security team to be dispatched to Moscow equipped with experimental X-ray equipment that could inspect construction elements without destroying any part of them. Even when other American teams were bringing new radiographic equipment to inspect structural columns, the Soviet workers mysteriously went on a two-week strike regarding their health hazards.

As the time went by and after several inspections from security expert teams, they discovered things that didn't belong where they were found based on the drawings. It was discovered that the Soviets had planted permanent eavesdropping systems in the actual structure of the building. They found cables in the concrete, interconnecting systems that were so sophisticated that couldn't be removed from the structural elements. They were placed in the steel and concrete columns, the beams, the precast floor slabs, and the sheer walls between the columns, making it impossible to remove. Electronic "packages" were found where steel reinforcement in the flooring should have been and resonating devices that allowed the Soviets to intercept all electronic communications along with conversations taking place in the building. The discovery of all the intercepting equipment was made even harder since the Soviets have placed decoys made to look like bugs and garbage from the construction process.

When the problems were thought to be irreparable, the US government shut down the project, leaving the eight-story building useless, infested with interception equipment planted by the Soviet construction workers.

7.5.6 *Portable X-ray Device*

Following the finding of a suspicious device by the nonlinear junction detector, a portable X-ray device may be used in order to "X-ray" the suspicious device, which cannot be examined otherwise. The cost is quite high, of course.

7.5.7 *Thermal Imaging Cameras*

Apart from X-rays, there is also another interesting "introspection" solution which has been used with great success for the past few years and that even helps locating devices that do not necessarily emit radio frequencies at that moment. It's the thermal imaging camera. A thermal imaging camera can detect subtle temperature differences generated during the operation of an electrical device (in our case, a bugging device). It thus indicates the location of a suspicious object from its thermal footprint. It is particularly useful in detecting hidden devices in other innocent objects, and, according to its sensitivity, the camera is even able to detect electronic devices on standby mode.

The example that follows is quite illuminating. Figure 7.10 shows the thermal picture of an air conditioner. The air conditioner in the Director's office was out of order for a long time due to failure (burnt fuse board). The figure obviously proves the existence of a heat source; following dismantling of the air conditioner, a malicious device was revealed, with a great battery supply, thanks to the wide and

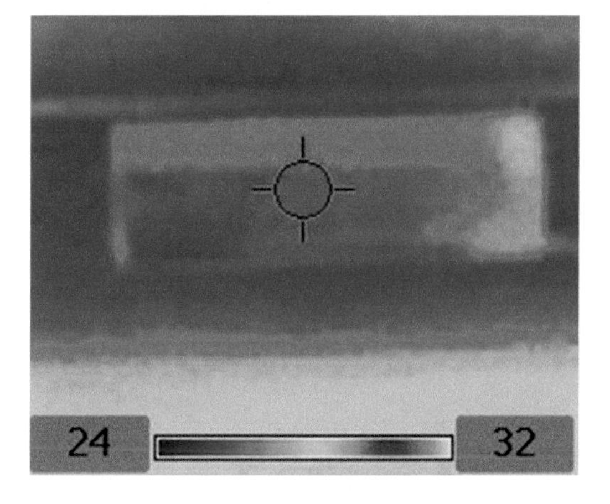

Fig. 7.10 Air conditioner thermal imaging. Notice the thermal footprint of the suspicious device on the *upper right corner*

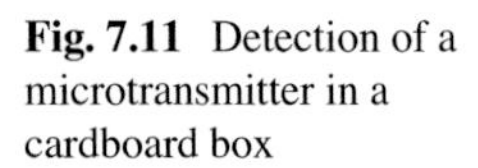
Fig. 7.11 Detection of a microtransmitter in a cardboard box

welcoming space in which it was placed. Furthermore, as this happened during the wintertime, the offender would have plenty of time to remove the device before the summer, when possibly a technician would be called to repair the air conditioner. A corresponding example is shown in Fig. 7.11, in which a microtransmitter has been placed in a cardboard box. Again the thermal footprint leads to its immediate identification.

7.5.8 *Infrared Detectors*

Despite the fact that the thermal camera measures the infrared emissions' heat, it is unable to locate monitoring devices that themselves emit infrared radiation (IR), because they use a different infrared band. Nevertheless, such devices can easily be detected using a CCD camera, whose IR filter has been removed. Some type of night vision devices may also be used, always paired with a filter to suppress the visible radiation letting only the infrared through. Finally, special heads can be attached to the frequency detectors, shifting and "translating" the infrared radiation to the radio-frequency band in which the detector operates. It must be noted that interception by infrared emission requires line of sight, and the suspicious device should therefore have been placed facing a window.

7.5.9 *Camera Lens Optical Detection Device*

Apart from a frequency receiver, which was mentioned earlier, a simple and cheap instrument (Fig. 7.12) can also be used to detect a hidden camera, exploiting the principles of optical reflection. A laser diode targets the area, and, if a camera lens

Fig. 7.12 Simple lens detection device

is spotted, the operator sees the reflection signal through a screen with a corresponding filter. Apparently, the laser beam is reflected on any other reflective surface, so the operator should thoroughly examine all findings.

7.5.10 Detection of Telecommunications Interceptions

7.5.10.1 General

Telecommunication equipment is obviously a favorite target of spies all over the world. In order to eavesdrop on them, several emitting devices may be used (powered by the phone line or by a separate source), as well as clip-on or passive, inductive devices. There is always the risk that ambient discussions in the entire area may be eavesdropped, not necessarily just the telephone conversation. It should also be noted that modern digital telephone exchanges provide the opportunity for lawful (or not) listening – in, i.e., without any kind of possible detection, but only by examination of the software running.

7.5.10.2 Clip-On Detectors

There are several devices for detecting tapped phones advertised in the amateur market (Fig. 7.13). What they usually do is to identify variations in voltage or line power. One could do the same with a simple multimeter. This method is effective only when someone connects a phone on parallel and picks up the handset, which usually happens in a residential installation. It can also detect a transmitter directly powered by the phone line and perhaps detect a primitive recording device which, due to its poor design, uses a lot of current from the line.

Every other eavesdropping device either uses an infinitesimal amount of current drawn from the line (based mainly on its own power source) or is inductively coupled; therefore, there is no detectable variation in the voltage or current. Besides, the line voltage by the telephony provider fluctuates during the day, depending on traffic and system load (it is well known that telephone exchanges supply power to the household phones, regardless of the main power network, which is why the classic, non-VoIP phones can still operate even during a power failure).

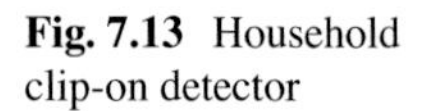

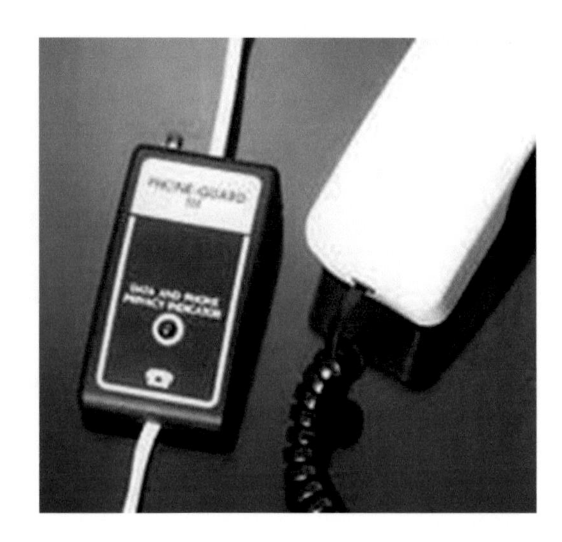

Fig. 7.13 Household clip-on detector

7.5.10.3 Telephone Testers

Equipment such as the one shown in Fig. 7.14 can test older, fully analog telephone exchanges. They were used to detect malicious devices which exploited technical features that do not exist anymore (e.g., the ability to establish an audio path during ringing and to activate a remote device sending tones during the ringing phase). The high voltage used in some of these tests would most probably destroy modern telephone exchanges before actually detecting any eavesdropping device. At the same time, such equipment could detect the existence of micro tape recorders, a museum piece by now, given the fully digital recorders of today.

Despite their antiquity, one of their functions can prove its usefulness even now: the one that performs automatic checks on multiple cable pairs. In a multipair cable, all possible pairs of wires need to be checked for possible intrusion devices and clip-ons. Even if the intrusion occurs with no physical circuit shortening but rather using a transformer or a capacitor, the tester can still emit tones to specific cable pairs and detect their presence in other pairs, thereby indicating the existence of some sort of tampering (or cross talk that still can be dangerous since it relays communications to other cable pairs). Another function amplifies the possible audio signal received from the cable pair, revealing the case where the telephone is tapped to receive the ambient sound of the surroundings, as a microphone.

In any case, the evolution of technology has led to more modern equipment, like the one shown in Fig. 7.15. This testing device is able to detect the presence of audio in telephone cables, as well as electrical abnormalities, which could be caused by the presence of a malicious device. For its operation, apart from the reflectometer, which we will examine later on, it also uses a more modern method, during which the signal reflection is inspected at different frequencies (frequency-domain reflectometer). It also uses the idea of the nonlinear junction probes, by inducing a microwave transmission in the line and by detecting possible harmonics denoting the presence of semiconductor devices along the line. Finally, it detects problems in VoIP-type communication, by incorporating the appropriate protocols.

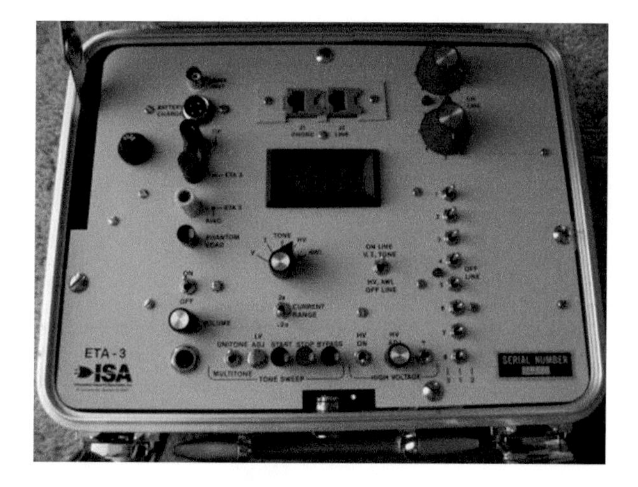

Fig. 7.14 Analog line tester

Fig. 7.15 Modern call line tester

7.5.10.4 Time-Domain Reflectometer (TDR)

This device detects abnormalities found in a wiring, thereby detecting devices that would be otherwise revealed only with a visual inspection. It may, for instance, reveal a clip-on cable leading to a second phone. In this case, and since there is no broadcasting signal, the radio-frequency detector receiver would not be able to identify the problem. Technically, the reflectometer may be described as a wiring "radar." It operates as follows: it emits a short voltage pulse across the wire and then records its reflectance. Each type of defect (short circuit, cable break, change of material, bending, twisting, etc.) changes the impedance at that particular point, and as such, the characteristics of the reflected pulse change too. Similarly, depending on the time it takes for the pulse to return, the time-domain reflectometer (TDR) can calculate the exact location of the problem on the wire. Indeed, by using the reflectometer, it is possible to detect the presence of a malicious device from hundred of meters away! It can also locate the exact point where something had previously

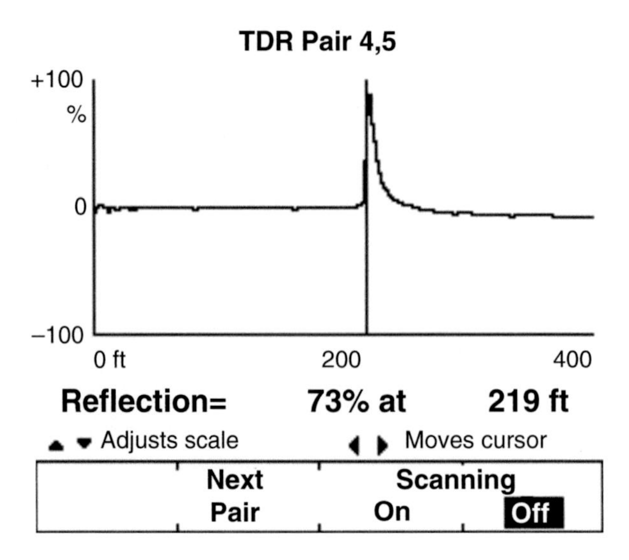

Fig. 7.16 Reflectometer screen

been connected and was later disconnected, leaving a wire "splicing" only. Figure 7.16 shows a screenshot of the display of machine TDR, which presents an example of a discontinuity at 219 feet distance in the pair 4,5 of a UTP cable.

As already mentioned, the latest version of such equipment is based on inducing radio-frequency signals rather than simple pulses, so the results can now be even more detailed. A problem fundamental to TDRs, however, still remains. Precisely, due to the long distance (kilometers) between our phones to the premises of a telecommunications provider, a TDR is likely to show dozens of wire abnormalities-defects. Many of them can very well simply be the results of work carried out by the provider (e.g., cable repair or routing change) in parts of the network that are not accessible to the user and owned by the provider. So the detection of devices installed along the cable becomes much more complicated outside the internal network. In any case, it is still difficult to examine inaccessible wire routing places, even in the internal network.

7.5.11 Conclusion

Leaving aside simple and amateur equipment, the supplies of a technical surveillance detecting professional, who respects himself and his profession, start from an initial investment of at least 50,000 euros. In more advanced cases, the total equipment value may well exceed 100,000 euros. Despite the considerable degree of automation, simply buying the equipment does not guarantee getting results. They must be accompanied with ongoing training and used by an experienced operator. In the following sections, we will analyze the precise methodology for performing the detection (site surveying, steps, tasks, etc.), as well as the process of finding the correct contractor to undertake a bug sweeping project (financial data, how to distinguish amateurs from professionals, etc.).

7.6 Methodology for Detection of Electronic Interception

Having dedicated so many pages on equipment, one can get the impression that the detection of interceptions consists only of scanning the area for traces of illegal transmission devices. The truth is, however, that this procedure is much more complicated and includes many other steps, which we are going to analyze in the following pages. The whole process would start with the initial contact with the counter-surveillance expert.

7.6.1 *Initial Contact*

The initial contact should take place outside the premises, where there is inkling that industrial spying occurs. No matter how trivial this error might seem, in the course of their panic, victims might contact the counter-surveillance expert the same telephones or areas that already "bugged." This move would of course inform the spies that they are being suspected, giving them the opportunity to remove the intercepting and monitoring devices before the check. It is therefore imperative to use a "clear" means of communication.

7.6.2 *Follow-Up*

After the initial contact, a meeting has to be arranged. This should be held in a carefully chosen or public place. Points that need to be discussed during this meeting are related to the concerns of the client, the indications, the sensitive issues, and the important information he tries to secure. The procedure that is going to be followed is decided, in the first place. A serious company will check with discretion the requests of the client for detection, and it might even suggest that there is no need for further investigation, protecting him from unnecessary expenses. During this meeting, many questions will be asked. These questions will allow the specialist to deduce whether there is an actual need for his services, which are the possible threats, and if this is a case of a legal counterintelligence operation or not. It might be the case that the client is nothing more than a criminal that tries to protect himself from legal surveillance by the authorities! Typical information and issues, that have to be settled before the check, can be sent by courier or discussed privately, and they include, among other things, the points in Table 7.1.

Table 7.1 Initial information

Detailed description of the area and the neighboring buildings, exact address and driving instructions. If other companies share the same building, then it is advisable to mention their names and activities
Information on dates, when constructing or other works (renovations, installations, etc.), are scheduled that will probably hinder the investigation
Full description of the company and its areas, offices with furniture and desks, their use and their surface. It is extremely important to note the existence of suspended ceilings and floors in every room. In that way, an initial time scale estimation that will be needed for the investigation can be given
Plots and layouts of the building in detail. Electricity supplies, data office, and furniture (desks, chairs, shelves, bookcases, tables, lockers, etc.) should be marked on the layouts. The type of plugs (single or dual) plays also an important role as it effectively doubles the necessary work
Type and number of telephones and PCs in every office
Type and number of fax machines
Type and number of photocopy machines
Type of private branch exchange (PBX)
Location of the central operator's console of the call center (usually at the reception desk)
Telephone calls and telephone connections (e.g., consulting the respective telco bills)
Telephone number for secure communication (e.g., use of special cryptophones)
Location of the main or intermediate distribution frames
Information and reports from other past investigations (if any) by other companies
Notes on recent restoration works, installation of new furniture, and delivery of new equipment (as it is quite possible that during this delivery, interception devices might have been installed)
IT and computer facilities (number of computers, servers, routers, switches, etc.)
Number and type of light fitters, ceiling, and floor lamps (fluorescent or filament lamps)
Number of light switches and dimmers and also air-conditioning and heating switches (e.g., thermostats)
Number of alarm sensors (e.g., motion detectors, door sensors, etc.) and smoke-fire sensors
Number of small decorative objects (clocks, calculators, hi-fi cd players, lighters, ashtrays, pots, signs, announcement/writing boards)
Number of ceiling and monitor speakers

7.6.3 Detection Planning and Choice of Areas

Taking into consideration all the previous information, a comprehensive overview will start over the plots and the drawings of the building, in order to set up the procedure to be followed during the detection. Parallel to that, the areas, where the operation will take place, have to be chosen. It is quite difficult for the investigation to include all the areas, not only in terms of cost but also in terms of required time too. There is a risk assessment involved, so as to rate and assess the security priorities. In this way, the investigation will have the ability to focus on the places where there is an actual danger of interception taking place. These areas will be examined carefully, in accordance with what we are going to describe in a few lines below, for the revelation of the points, where interception and monitoring devices have possibly been installed.

Areas, which will obviously have to be examined include the meeting-conference rooms, other smaller meeting rooms for discussions with clients, the executive and secretary offices, the research and development departments (if any), the legal department, and, last but not the least, the main distribution frames and telephony lines. Respectively, equipment in these areas has to be examined thoroughly, such as telephones, fax and photocopy machines, printers, scanners, etc. In addition, neighboring areas to the ones above have to be examined also, for example, storage places, floor distribution frames, waiting rooms, corridors, file-storage rooms, and libraries. Especially for the storage places, a quick visual check is advisable, as it is quite possible to find large electronic appliances, like transponders, which receive a weak signal from neighboring offices, where interception devices are installed, amplify it, and transmit it much further away, so as to reach the spies. As mentioned in a previous chapter, the main "bug" can be placed on the target's office, while the amplifier repeater can be placed in a nearby office offering more space.

If the executive offices and conference room prove to be clean, then it is rather difficult to expect that an intercepting device is installed somewhere else (e.g., at the reception, at the restaurant, at rest rooms, etc.). However, what we usually forget is that the executives do not work only within the company premises. Important meetings and discussions take place outside the working environment, be it at home, in a car, or even on a boat! Therefore, a check should be scheduled for these areas too, starting from the home and the car of the executives.

7.6.4 Identification of the Areas and Vulnerability Analysis

External and internal identification survey of the areas will follow for the best possible preparation for the detection procedure. All the collected data, in combination with the previous information, will lead to the development of a threat-model and a vulnerability analysis. During the identification of the external areas, the expert examines the surroundings which appear to be suitable for interception (e.g., neighboring buildings, constructions, parking lot, etc.). He also looks for places where a receiver may be placed, receiving signals from the inner part of the building. Respectively, he examines the entrance halls which give access to fraudsters and to other malicious competitors. Usually, this is programmed for the day before the check, so as not to raise any suspicions.

Next, during the identification of the internal areas, a touring takes place in the company with discretion in order not to cause suspicions and questions, so as to have a clear picture, according to what we described in the Sect. 7.6.3 (Detection Planning). Building elements are marked (type and wall insulation, existence of suspended ceilings and floors, skylights, wells, wiring electrical connections, etc.). After that, the expert tracks down the sensitive points and points that are susceptible for free area-natural sound transmission, and auditory information leaks from one place to the other. Furthermore, he notes down possible spots where there "monitoring locations" can be accommodated. For this reason, he lists the furniture and the decorative objects in the room. Closing, he makes an assessment of the physical security system (alarms, locks, etc.) and its weaknesses.

7.6.5 Transmission Detection

At this point, the actual detection procedure itself starts. For tactical reasons, it is suggested to start with a passive control using electronic equipment, so that annoying noises and sounds that could alert any possible eavesdropper won't be heard. Frequency scanning for wireless transmissions, detection of infrared and laser emissions, as well as detection of optical modulation interception devices are carried out at this step. The electronic detection equipment is adjusted accordingly, so that it is not using any audio signals but only optical indications (otherwise, headphones can also be used). Soft ambient music or another continuous acoustic background (e.g., an ordinary discussion) equally helps both in the detection and in the camouflage, as not to alert a possible spy. This introductory procedure of detection could be brief and concise, and it reveals directly the low-priced and unprofessional monitoring devices. But of course, in urban environments with intense electromagnetic activity, the time required for the frequency scanning is multiplied, since there are dozens of signals that have to be assessed, with the majority of them being legitimate.

Another important issue concerns the timing of the investigation. The detection process should take place not only during working hours but also beyond them. Usually, the biggest part of the procedure is better to be performed out of the working hours, so as not to raise suspicions to the personnel or to the any possible insider spies. However, the interception devices could be configured to transmit only during working hours in order to consume less power allowing for an extended working time. For this reason, part of the detection procedure must definitely be performed during working hours.

The truth is that it is not easy to keep all this activity classified. This is even more the case, when there are hints that the spy is an insider. As such, the operation must be run extremely carefully. In order to make that happen, the client can make up a scenario to camouflage the detection team when entering the building (they can be, e.g., presented as a lab team measuring air pollution) or gather the personnel in other places with the excuse of an evacuation exercise or an urgent meeting that came up.

7.6.6 Electric and Electronic Checks

Apart from the transmission detection, a series of electrical and electronic checks need to take place. They include activities of Table 7.2.

Table 7.2 Electric and electronic checks

Check all the building wiring (telephones, data, electricity, switches, alarms, fire alarms) using the TDR instrument
Detection of telephone lines and checking of the distribution frame wiring connections
Detection of provisions transmitting heat, hidden in other objects, using thermal cameras
Scanning using NLJD equipment for the detection of non-active or even decommissioned electronic bugs
X-ray analysis
Detection of microphones with the method of acoustic feedback (this is the last stage of the procedure, because due to the audio feedback, a revealing sound is produced. This could put in jeopardy the operation, as it would inform the eavesdropper that he is being discovered)

7.6.7 Simple Visual Check

Undoubtedly, the importance of electronic appliances and equipment during the detection is determinant; however, they should not exclude a plain visual check. Despite their potentials, most of them cannot trace a device which is not powered (e.g., when run out of battery) or a device that is based on a passive operation (e.g., a simple wired microphone). Indeed, sometimes a "tricky" wire has to be followed all its way through the building (suspended or not ceilings, skylights, wells, basements, etc.). Even the NLJD equipment has important restrictions regarding the distance requirements for the detection of electronic devices. Their operation may also prove unsatisfactory because of other ordinary electronic equipment (e.g., a television set, a computer) in the room.

So, a plain visual check still remains necessary. Counter-surveillance companies that only trust detection equipment without applying a thorough visual simple check should be avoided. For sure, a simple visual check lacks the prestige compared to a super expensive detection device. A visual check is a "dirty" job, and you have to roll up your sleeves; use humble tools; kneel and search with a torch under the desks, the sofas, and the chairs; climb stairs and check suspended ceilings; and so on. Actually, many companies follow a specific procedure, and they have a list of things to be done, in order to maximize their results without leaving any object unexamined.

We have already seen in a previous chapter how miniature the recording devices and the microphones can be. This allows them to be embedded into any other object available in the room. Experience shows that there exist particular spots Trojan horses that support the installation of such bugs. It goes without saying that some of these things are self-evident, while others ask for a "wild imagination" (or experience accordingly), in order to be investigated. In any case, the ingenuity of the spies leads to new places being exploited in order to place interception devices. As a result, the check of the area should not be limited to the "popular" spots. An indicative, but under no circumstances, exclusive list of the places that should be taken into consideration for examination is given in Table 7.3.

All of the abovementioned places must be thoroughly examined. Obviously, the people in charge have to disassemble and deconstruct all the electric and electronic

Table 7.3 List of places possibly hiding a "bug"

Under the tables, chairs, armchairs, sofas, and other furniture
The backside of bookcases, boards, filing drawers
In between books or even inside the spine of books
In between files in filing drawers
Closets and cupboards
Decorative objects, souvenirs, paintings, sculptures, pots, and promotional gifts
Ink pots, presse-papier, ashtrays, pencil boxes
Power supplies and data network (wall and floor installations)
Suspended ceilings and floors
Floor standing lamps and desk lamps
Office ceiling lighting
Electric and electronic appliances such as calculators, radios, decorative lamps, and wall and table clocks
Telephones
Office equipment (e.g., fax and photocopy machines, printers, scanners, etc.)
Building elements (door and window framings, plasterboards, etc.)
Behind and in between central heating radiators, boilers, cooling-heating air ducts
Smoke and fire sensors
Alarm sensors
Light switches, air-conditioning and heating thermostats

appliances, unscrew all the lamps, etc. Inevitably, this is a very noisy task that could alert the eavesdropper.

Respectively, the walls, the floor, and the ceiling have to be examined, to uncover recent works that can have left signs of slight bores, dust, paint, etc. They can very well be signs of the process of hiding an interception device. The suspended ceiling and floor both have to be lifted. Even the most inaccessible spots can be examined with the use of specific borescopes. During the investigation, quite often keys are found. In this case, it is necessary to check if the keys fit to locked drawers in the area and write it down. There is no need to lock the drawers if you keep the keys in the same room! All the spy has to do is look for the keys.

7.6.8 Telecommunications Check

The telecommunication equipment check, due to some specific peculiarities, deserves special attention. Especially for the private branch exchanges (PBX), extra time and exceptional experience are required for their check, because there are dozens of different types, which constitute (at least traditionally speaking, before the VoIP era) proprietary development platforms, with a difficult environment to manage them. In order to check and audit their settings and make sure that any surveillance services are deactivated, it is absolutely taken for granted that an expert should be hired, in addition to the rest of the team that will handle the "standard" detection

Table 7.4 Telecommunication equipment check

A simple visual examination of the area, of the distribution frames, of the wiring, and of the telephone sets for signs of interception devices installed
Checking of the plugs using the TDR for revealing tapping-splicing, as well as wired interception devices
External penetration testing of the PBX
Internal penetration testing of the PBX
Disabling of the remote tele-maintenance equipment, so as to prevent non-authorized changes-modifications of the settings at the PBX
Documentation on wiring and distribution frames for a future reference
Visual check of the FCT (equipment connecting the cellular network telephony with the call center for cheaper calls to mobile phones)

Table 7.5 Optional actions for checking telecommunication equipment

Check the settings, the services, the recordings (logs) at the PBX
Check the tele-maintenance settings
Enable full logging functionality
Check the call detail records for any patterns of fraud

of interceptions in the company. This expert can also secure the PBX against the possibility of abuse and overcharging by fraudsters that can attack it.

Apart from all the previous actions taken that can detect surveillance devices in telephones and wires, a series of additional checks should be practiced at this stage, which include, among other things, the points included in Table 7.4.

Optional actions depending on the availability of software and hardware access codes are included in Table 7.5.

Now, moving to the public telephone network, it is really difficult to secure the confidentiality of communications only by closely looking at the equipment and installation of the client. This happens for a simple reason; the connection of the public telephone network passes through hundreds or even thousands of meters outside the company, in order to reach the telephone exchange. At the same time, this could make it difficult for the spy to set up a device outside the company (yet, working in collaboration with a corrupted telco provider employee, this could be easier for him). Thus, a cryptophone is necessary, provided that the counterpart has a compatible model.

7.6.9 Computing and Network Infrastructure Check

Security of computing systems is a huge chapter itself that cannot be analyzed in this book. This task is assigned to specialized companies in external and internal penetrating testing, which practice comprehensive security checks. We also have to mention that traditionally, companies active in the fields of interception detection and telecommunication monitoring were not highly qualified for the examination of IT infrastructure. Many of these, even today, do not offer the suitable specialization, so this has to be clarified from the beginning. In any case, a full detection project should forecast in advance to secure a budget for IT security audit, as it is quite possible that industrial spying may occur entirely via the computing and network infrastructure, without any other technical intervention or use of interception devices.

7.6.10 Findings, Deliverables, and Next Steps

In case that something suspicious is discovered, documentation of the finding, identification, and analysis should follow, and then authorities should be informed. It is also probable that a forensic analysis is requested in order to extract exhibits to be used in the court of law. If a transmission device is uncovered, it is important to keep it active, so that the spy will not realize that he has been exposed. On the contrary, ordinary discussions should continue to take place in the area, in order to make a story with false information that will lead to framing and the revelation of the spy.

A complete report has to be handed in, as soon as the operation is terminated. The report should describe the findings (if any), the equipment used, and the procedure that was followed. In addition, it is important to mention the places that were under surveillance, and the spots found susceptible or sensitive to interception, so as to protect the company from future attempts. Counseling about technical countermeasures and security equipment, in terms of practice and information, is equally important. The report has to be comprehensible to the ordinary user, avoiding superfluous and specialized terminology. However, a technical appendix is necessary which will be evaluated appropriately by experts and can be used for future reference. The report is also suggested to be analyzed over a presentation that will bring about ideas, corrective actions, and further steps.

A very basic point that must be mentioned is that every technical surveillance detection project is a snapshot in time of the present situation. In other words, an interception device can be installed only a few minutes after the expert team finishes the investigation and leaves the building. Therefore, it is highly recommended that frequent repetitive checks take place. Usually, brief and basic detections are performed at regular intervals, until a full detection operation takes place, according to what we have described so far.

7.7 Preventive Measures and Area Protection

Preventive measures start with the physical security and the access control. This renders any illegal entrance difficult and minimizes the attempts to install "bugs." When the company is relatively small, it is easy to track down any unknown incoming people. In big companies though, a reception desk is necessary, and possibly an access control system would be a good idea.

On the other hand, even strict security measures can be circumvented not only by determined spies but also by the working staff that constitutes an "inside" risk, voluntarily either not. Therefore, it is important to restrict the flexibility of employees to intervene in working areas that they have no reason to be (e.g., an accountant visiting the research department). Especially when it comes to the "inside threat," apart from the technical detection equipment for data leakage and the forbidden use of removable data storage appliances, regular checks should take place.

Moreover, visual checks can be practiced by the company personnel itself, following a basic training. Even without equipment, this method is a very good idea for a baseline protection from basic threats, until professionals take up the case. Employees can actively help searching for wires and cables, checking new furniture, wearing on walls and floors, etc. according to the instructions we have described few pages back.

Regardless of the technical examinations, taking precautions is also important. Table 7.6 presents some evident prevention measures that demand for special attention.

If a company starts implementing these suggestions and repeats checks and audits at regular intervals, in combination with good practices, information security management systems (such as ISO 27001), and educational programs, then it takes most of the necessary precautions against industrial spying.

Table 7.6 Preventive measures

The personnel should be reliable and satisfied with the working conditions in the company, and they should be trained, so as to realize the importance of confidential information
Except for the company secrets, it is also important to secure everyday simple and ordinary information necessary for the company's operation, such as salary slips and contracts that should be locked in drawers. If an employee realizes that one of his colleagues at the same department is better paid, it is highly possible to adopt a negative attitude toward the company and turn into a threat
At the same wavelength, when the employees leave the office, they should lock confidential documents in their drawers
Preventive interceptions and audits must also be scheduled before important meetings, especially in the rooms where these meetings will be held
It is also recommended to install jamming devices in sensitive areas and privacy assurance systems for communications (e.g., cryptophones)
Sensitive objects can be marked with markers of "invisible" ink (it can be seen only under ultraviolet light). In this way, one can see if someone has touched the objects during his absence. A very typical example of this method is the marking of the telephone screws, by drawing a line. If someone unbolts a screw and then tries to bolt it back, he will miss the exact previous spot. Thus, a marking will be created (as shown in Fig. 7.17). In the upper part of the picture, we have marked with light blue the invisible ink, in order to make it clear to the reader. When the screw is tampered with, its position has changed, and this is evident as it can be seen under ultraviolet light. A more "aggressive" method is the use of color pigment, which leaves marks on the hands of "the curious," in case there are suspicions of an insider

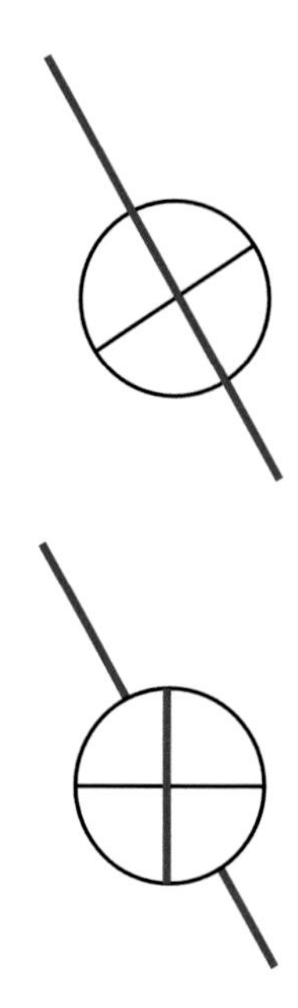

Fig. 7.17 On the *top* is the "invisible" *line* under the screw, as it can be seen under ultraviolet light. At the *bottom*, it is obvious that the screw has turned after bolting and unbolting

But what happens if something wrong is found out in the company? Keeping calm is the first step. Do not remove the device. The finding has to remain secret even for the people in the company. Also, the place has to be secured so that no further access should be made possible. At the same time, trivial discussions must be held, so that the eavesdropper will not realize that his device has been discovered. After following all of the above, the client will carefully come into contact with a specialized company in counter surveillance and interception detection, which will guide him to the next steps. At this point, we will present how to make the right choice for choosing a contractor for a counter-surveillance project.

7.8 Choosing the Right Contractor

7.8.1 Introduction

Fear (sometimes groundless) for industrial spying can lead many executives to be exploited by fraudsters. In order to safeguard a company against industrial spying, there are some parameters that have to be taken into consideration, including risks and countermeasures. A risk assessment can be of great help. The company evaluates the importance of information it poses, the consequences of its loss, and the possibility of interception. In cases of interception, the key role that lies is the third entity. Who is interested in obtaining this information and the resources the company has?

Naturally, the "opponents" are not "weighted" with the same importance. The first thing one should think about is "What kind of sensitive information does my company possess that a third entity would take the risk to illegally access?" Once we realize that there is such a risk or we already have indications of data leakage, then we should move on to detection procedures. However, monitoring and interception detection is not a panacea. Regardless of the findings, all the fundamental security principles, as well as the preventive measures described, have to be followed.

The detection of interceptions is a technical task that does not necessarily fall into the responsibilities of the authorities. Authorities take on from the moment a device is found, so that legal procedures have to be initiated (if requested). On the other hand, a company which makes an attempt to detect interception devices only depending on its own resources, without requesting expert's help, is condemned to failure. It is quite possible for someone to discover on his own an interception device, but this would be limited to cases of cheap devices used by amateurs. When we deal with serious cases of industrial espionage, then the contribution of experts is the only way to go forward, since only experts are well trained, experienced with updated and high-caliber equipment. These qualities make them irreplaceable.

7.8.2 Categories

Unfortunately, for such an important field, followed sometimes by "paranoid" tendencies, a large percentage of the companies offering counter-surveillance services are neither sufficiently trained nor possess the appropriate equipment. There are five basic categories:

Fraudsters People who actually have no connection with the subject, but they are capable of putting on a show pretending that they are searching for "bugs." A usual technique, among others, is the actual finding of an interception device, just like how professionals do. There is a minor detail; however, the device had never existed in the first place. They had brought them along with their visit there! As a consequence, they present to the client a guaranteed success. These devious people can sometimes be so brazen that the circuits they demonstrate are not real "bugs" but simply electronic accessories!

Sellers Beware of those who claim that they always find something. Their intention is to sell a useless and unprofessional detection device to you. Under no circumstances should you fall victim to these people who advertise "automatic" and "exclusive" equipment, which can detect "any kind of threat." Besides that, even if it is in fact a reliable, expensive, and professional device, the results depend on the operator, since proper training plays a very significant role. Indeed, think about this: if you buy a guitar, does that mean that you have right away learnt how to play music?

Amateurs Without training, using low-cost equipment just bought online or from an electronics shop. These may have actually fallen victims of the previous category of the sellers. Some of them may truly believe that they do a good job, but most of them are aware of their ignorance, and for this reason they will avoid answering your questions and will also avoid analyzing their work and their deliverables. Their services are more or less useless, and usually the cost of their visit is low. Of course we should not omit amateurs, who charge their services the same rate with the professionals. They should fall into the category of "fraudsters."

Semiprofessionals They are trained and they have invested in some basic medium-cost equipment (few thousand euros). They might mean well, but good intentions only are not enough for a demanding job like this. Even if they have invested in high-caliber equipment, lack of training and experience will turn out a big disadvantage. Falling into the two previous categories, we have companies installing burglary alarm systems, computer networks, and VIP security services and companies of private detectives. Under no circumstances at all should these people be called "professionals." Especially, the detectives are trained in actually installing interception devices rather than finding ones! As such, they are not considered a good choice.

Professionals They are characterized by substantial and comprehensive training, having attended dozens of seminars with a wide range of technical knowledge, a solid background, and a professional experience. They use very expensive and modern equipment (tens of thousands of euros). Furthermore, they have excellent knowledge of "know-how" regarding interception devices. The academic studies are not necessary, but a degree can be regarded as an extra asset.

Reports and recommendations from satisfied clients should always be welcomed. But, sometimes the job demands secrecy, even after the completion of the investigation, so recommendations may not always be available. In this case, the interested client could ask for a typical report sample (usually deliverables, with any confidential information removed), so as to get an idea and evaluate the quality of the services to be offered.

Having a look over the report, the client can ask the right questions to the candidate for this project, in order to check (at least to a minimum level) his knowledge. Fraudsters will not be able to give a reasonable answer, even to simplest questions, a sign that something is wrong. Just because they use incomprehensible technical terminology, it does not mean that they know the subject they talk about. On the contrary, you should insist that they speak to you in a simple language, avoiding unnecessary terminology and technical data that lead to confusion. Ask them about their training and the seminars they have attended. For example, "when was the last seminar you attended?" You should always ask for certificates on their experience and on their training. Do not accept as an excuse that the techniques and the equipment they use are strictly confidential. A real professional can answer to all these questions without fear, in plain words, explaining his work so that he can be clearly understood. Eventually, the more informed the client, the better choice he makes.

7.8.3 What Should Be Avoided

Table 7.7 presents typical categories of companies and people, who are involved in the field but have to be avoided.

7.8.4 Economic Background

Having chosen the right contractor for the counter-surveillance project, we should bear in mind the financial factor. The most important rule, in terms to economic background, is that the cost of the detection should be lower than the value of the information we are trying to safeguard and, of course, lower than the amount we can afford. Clients who discovered interception devices that could cost them millions of euros in profit losses rarely complain about a surveillance detection project costing tens of thousands of euros. In any case, the final financial offer depends on many other factors as well, as indicated in Table 7.8.

The financial offer can also be analyzed and charged based on units, for example, detection cost per square meter, per telephone appliance, per computer, etc. Unfortunately, the rates cannot be regarded as a criterion of professionalism because

Table 7.7 Avoid

Companies which mention that they can discover everything: these statements are indicative of amateurs or sellers. A real professional knows how difficult the detection of advanced interception devices is and does not make such irresponsible statements
People with criminal record: these might be fraudsters who intend to deceive you by presenting a "fake" detection. This is the best case scenario, as the worst one might be that the same people could actually commit industrial spying against your company, under the cover of detection
Excessive advertising and public relations: both might be signs of lack of expertise, and they simply denote a show-off
People who introduce themselves as "former" spies: since by definition, they were involved to "classified" cases, they cannot reveal to you more evidence. So what is the point of this statement?
Service contractors: ask if the person you are talking to will undertake the investigation himself. There are companies which use qualified "sellers," in order to convince their clients to hire them. What happens in reality is that a group of subcontractors or unskilled and unpracticed technicians show up
Experts who come from the security services: despite their knowledge and training, there is always the risk that they are left behind in terms of technological evolution, and they are not updated on current developments. It is also quite possible that they used to work on a totally different case that did not have anything to do with detection of interceptions. Unlike the previous categories, it is highly possible to find remarkable professionals in this group. So, having in mind the questions in the previous paragraph as a helping guide for the proper choice, one may end up with a good cooperation after all

Table 7.8 Factors defining the cost of detection

Factors defining the cost of detection
Transfer and accommodation expenses of the team
Brand name and reputation of the company
Company's investment in equipment and training programs
General costs and profit margin
Market rates
Estimation of the time needed. A prior visit or at least a study of the layouts and plots of the place to be examined would be of great help to an accurate time planning. This is important, because if time estimation is wrong, then we run the risk that the investigator will lower his level of work, in order to reset the profit to his initial standard. Obviously, "cramped" offices, with dozens of objects and suspended ceilings, need much more time than a conference room with very few objects and a proper ceiling
Job hazards (e.g., involvement with state secrets)
Possibility of legal sanctions in case something goes wrong (e.g., cause of damage)

many skilled fraudsters charge the same with the professionals, counting on client's ignorance. Typical rates for a small company with few offices start from some thousands of euro for a one-day detection and escalate according to the size of the company and the factors we have just mentioned. A completely "basic" detection, even though it does not offer great security, is quite useful to occur at regular intervals, when the level of risk based on the relative analysis is low.

Another "sensitive" point that should be mentioned, in relation to the economic background, is that many times it is asked from the client, especially from new ones, to pay in advance the whole amount, before the completion of the investigation. Indeed, most of the time, there is nothing dangerous found in the company, and this tactic protects the detection companies from clients who are not willing to pay, once they realize that they do not encounter industrial espionage problems. Needless to say, there are a great number of clients who have a wrong way of thinking, i.e., "what is the reason for paying, since nothing was found?" In any case, the detection company might ask for more details about the client, in order to make sure that he is not a competitor who tries to find out the rates by pretexting to be a client!

We have to point out once again that when the value of information that the company wants to safeguard is much higher than the detection cost, "bargaining" should be avoided at all costs.

Printed in the United States
By Bookmasters